ELECTRONICS AND CIRCUIT DESIGN MADE EASY

KERWIN MATHEW

ELECTRONICS AND CIRCUIT DESIGN MADE EASY

PREFACE

This book has been written with practically all types of readers in mind: laymen to engineers. It is presented in an "instructional" style, precise and straight to the point, with no "beating about the bush". With this book, it is not necessary to make notes, as points are clearly and logically stated, point by point. The book is so simple in form that the reader should have little difficulty in mastering some circuit design technique - he should be able to design some simple circuits at least.

The book covers both the theoretical and practical aspects of electronics and circuit design and would act as a manual and reference guide for the student of electronics and the practicing engineer.

Kerwin Mathew, Ph.D., PE, CMfgT, CPM

CONTENTS

1 A FEW BASICS

Ohm's Law

1.1   Ohm's law states that:

Voltage = Resistance x Inductance

Resistors In Series

2.1   The following figure shows two resistors connected in series:-

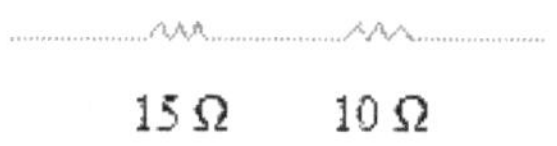

15 Ω   10 Ω

2.1.1 Their total resistance is as follows:-

$R_T = R_1 + R_2 = 15$ ohms $+ 10$ ohms $= 25$ ohms

2.2   The total resistance is often termed the "equivalent series resistance" - $R_{eq.}$

Resistors In Parallel

3.1   The following figure shows two resistors connected in parallel:-

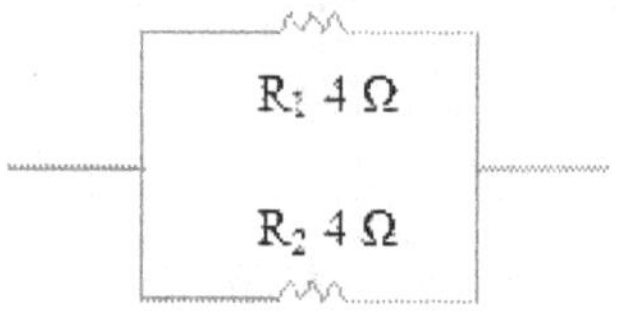

R_1 4 Ω

R_2 4 Ω

3.1.1 The total resistance in the above circuit is as follows:-
       $1/R_T = 1/R_1 + 1/R_2 = ¼ + ¼ = 0.5$ ohm

3.2   This resistance is often called the "equivalent parallel resistance".

3.3   The formula for three resistors in parallel is as follows:-

$1/R_T = 1/R_1 + 1/R_2 + 1/R_3$

Power

4.1   Current flowing through a resistor dissipates power, which usually takes the form of heat.

4.2   The expression for power is "watts".

4.3   The formula for power is as follows:-
       $P = VI$ or $P = I^2R$ or $P = V^2/R$

The Voltage Divider

5.1   A circuit diagram of the voltage divider is shown below:-

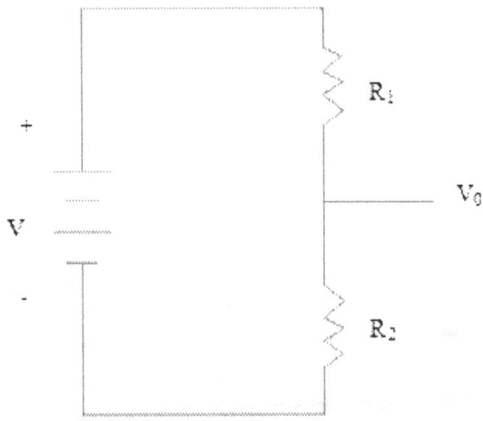

5.2   The voltage divider is the basis for many important theoretical and practical ideas throughout the whole field of electronics.

5.3   The important point in the above circuit is V_0, which is the <u>voltage drop</u> across R_2.

5.4   The formula for this voltage drop is as follows:-

       $$V_0 = V \times R_2/(R_1 + R_2)$$

5.5 $R_1 + R_2 = R_T$, the actual total resistance of the circuit.

The Circuit Divider

6.1   The circuit below shows the current splitting or dividing between the two resistors which are placed in parallel:-

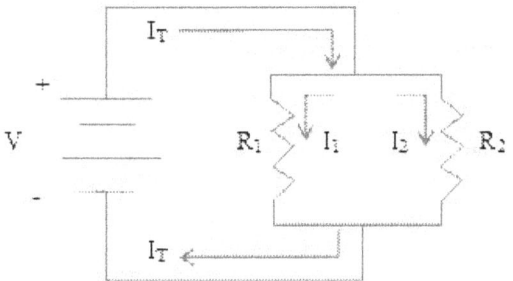

6.1.1 I_T splits into two individual currents, I_1 and I_2.
6.1.2 I_1 and I_2 then recombine to form I_T.
6.1.3 For the above circuit, the following equations are true:-
    [1] $V = R_1 I_1$
    [2] $V = R_2 I_2$
    [3] $R_1 I_1 = R_2 I_2$
    [4] $I_1 / I_2 = R_2 / R_1$ (Here, the current divides in the "inverse ratio of the resistance values".)

Capacitors In DC Circuits

7.1 In electronics, capacitors are extensively used.
7.2 Their main use is with AC signals.
7.3 However, there are certain specific areas of DC where they are very important.
7.4 The main use of capacitors in DC circuits is to become charged and hold the charge.
7.5 In the diagram below, the capacitor will charge when the switch is closed:-

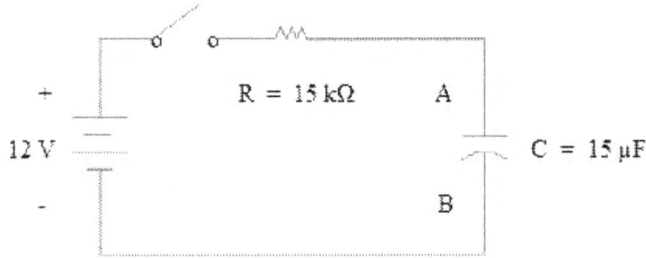

7.5.1   In the above circuit, the final voltage to which the capacitor will charge is 12 V.

7.5.2   The capacitor will charge up to the voltage that would appear across an open circuit that is located at the same place where the capacitor is located.

7.5.3   The formula for the <u>time constant</u> of this kind of circuit is as follows:-

$$T = RC$$

7.5.4   The time constant for the above circuit is:-

$$T = 15\,k\Omega \times 15\,\mu F = 0.225\ second$$

7.5.5   The time it takes the capacitor to reach 12 V in the above circuit is approximately 5 time constants, or about 1.125 seconds.

7.5.6   In one time constant the capacitor charges to 63% of the final voltage, or about 7.56 V.

7.5.7   The capacitor will be uncharged before the switch is closed.

7.5.8   When uncharged or discharged, the capacitor has the same voltage on both plates.

7.5.9   In the above circuit, before the switch is closed, both the voltage on plate A and plate B will be at 0 V if the capacitor is totally discharged.

7.5.10   In the above circuit, when the switch is closed, the voltage on plate A will rise towards 12 V.

7.5.11   In the above circuit, the voltage on plate A after one time constant will be about 7.56 V.

8.1   Capacitors can be connected in parallel as shown below:-

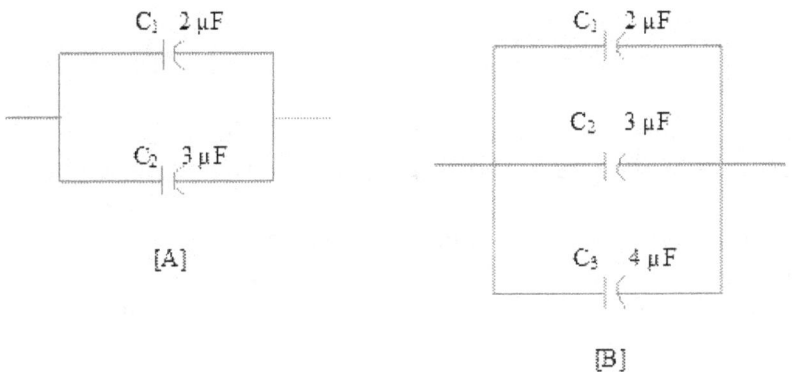

8.1.1   The formula for total capacitance is as follows:-

$$C_T = C_1 + C_2 + C_3 + \ldots + C_N$$

8.1.2   The total capacitance in figure [A] is as follows:-

$$C_T = 2 + 3 = 5\,\mu F$$

8.1.3   The total capacitance in figure [B] is as follows:-

$$C_T = 2 + 3 + 4 = 9\,\mu F$$

8.2   The total capacitance for capacitors connected in parallel is obtained by simply adding the capacitor values.

9.1   Capacitors can be connected in series, as shown below:-

9.1.1   The formula for total capacitance is as follows:-

9.1.2   In the above diagram, the total capacitance is as follows:-

$1/C_T = 1/2 + 1/3 = 5/6$

$\therefore C_T = 6/5\ \mu F = 1\ 1/5\ \mu F$

Switches

10.1   A device that completes or breaks a circuit is known as a mechanical switch.

10.2   The mechanical switch is most commonly used in applying power to turn a device on or off.

10.3   A switch can allow a signal to pass from one point to another, prevent its passage, or route a signal to one of several points.

10.4   There are two types of switches, the on-off switch or <u>single pole single throw</u> switch, and the <u>single pole double throw</u> switch.

10.5   Below are circuit symbols of the two switches:-

switch   ON-OFF switch   Single pole double throw or SPDT

in the OFF position

10.6   A CLOSED (or ON) switch has the total circuit current flowing through it, and there is <u>no</u> voltage drop across its terminals.

10.7   An OPEN (or OFF) switch has <u>no</u> current flowing through it, and the full circuit voltage appears across its terminals.

10.8   Below are circuits showing the operation of the single pole single throw switch (figure A) and the single pole double throw switch (figure B):-

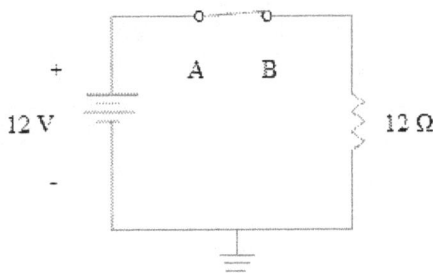

figure A

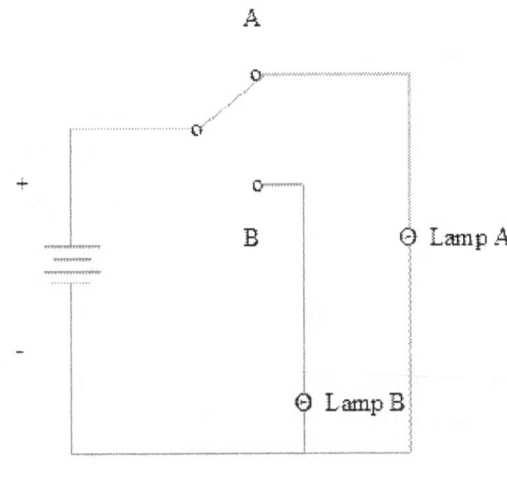

figure B

Review

11.1   The basic electrical circuit comprises of a source (voltage), a load (resistance) and a path (conductor or wire).

11.2   The voltage represents a charge difference.

11.3   If the circuit is complete, electrons will flow in what we call current flow. The resistance offers opposition to current flow.

11.4   Ohm's Law states the following:-

$$I = V/R$$

11.5   The total resistance for resistors connected in series is given by the following formula:-

$$R_T = R_1 + R_2 + \dots + R_N$$

11.6   The total resistance for resistors connected in parallel is given by the following formula:-

$$1/R_T = 1/R_1 + 1/R_2 + \dots + 1/R_N$$

11.7   The following equation gives the power delivered by a source:-

$$P = VI$$

11.8   The following gives the power dissipated by a resistance:-

$$P = I^2R = V^2/R = VI$$

11.9   If the total applied voltage, E, is known, the voltage across one resistor in a series/string of resistors is obtained from the following voltage divider formula:-

$$V_1 = ER_1/R_T$$

11.10 If the total current is known, the current through one resistor in a two resistor parallel circuit is obtained from the following current divider formula:-

$$I_1 = I_TR_2/(R_1 + R_2)$$

11.11 The voltage drops in a series circuit is related to the total applied voltage by Kirchhoff's voltage law (KCL), which is as follows:-

$$E \text{ or } V_T = V_1 + V_2 + \dots + V_N$$

11.12 Regarding the currents at a junction in a circuit, Kirchhoff's current law (KCL) states that the sum of the input currents equals the sum of the output currents. For a simple parallel circuit, the currents at a junction in the circuit are obtained from the following formula:-

$$I_T = I_1 + I_2 + \ldots + I_N$$

where I_T is the input current.

11.13 In a circuit, a switch is the control device that directs the flow of current or in many cases allows the current to flow.

11.14 In a circuit, capacitors are used to store electric charge.

11.15 Capacitors allow current or voltage to change at a controlled pace. The circuit time constant is obtained from the following formula:-

$$T = RC$$

11.16 The values for current and voltage would have reached 63% of their final values at <u>one</u> time constant in an RC circuit.

11.17 The values for current and voltage would have reached their final values at <u>five</u> time constants.

11.18 The following formula shows how capacitors connected in parallel are added to obtain a total value:-

$$C_T = C_1 + C_2 + \ldots + C_N$$

11.19 To find the total capacitance, capacitors in series are treated the same as capacitors in parallel.

11.20 For capacitors connected in series the total capacitance is obtained from the following formula:-

$$1/C_T = 1/C_1 + 1/C_2 + \ldots + 1/C_N$$

2 DIODES

1.1   The main feature of the diode is that it conducts electricity in one direction only.

1.2   The first vacuum tube was a diode. It was also known as a rectifier.

1.3   The modern diode is a semiconductor device.

1.4   It is used in all applications where the older vacuum tube diode was used.

1.5   It has the advantages of being much smaller, easier to use, and less expensive.

1.6   The term "semiconductor" describes a class of devices whose properties do not permit them to be classified with either conductors or insulators.

1.7   Under the correct conditions semiconductors will conduct an electric current in a well defined and controlled manner.

1.8   Semiconductors are the basic material of all modern electronic circuits.

1.9   The semiconductor diode should be studied as an introduction to the transistor.

2.1   Both silicon and germanium are semiconductor materials.

2.2   Both silicon and germanium are used in the manufacture of diodes, transistors and other components.

2.3   Both are refined to an extreme level of purity.

2.4   Minute, controlled amounts of a specific impurity are then added.

2.5   Depending on which impurity is added, the silicon or germanium can be regarded as N or P material.

2.6   The electric current in semiconductor material is made up of two types of charge carriers.

2.7   In N material the majority carriers are negative in charge (electrons), while the minority carriers are positive in charge (called holes).

2.8   In P material the reverse is true.

2.9   When a piece of N silicon is joined to a piece of P silicon, a diode junction is formed.

2.10   This is often called a PN junction.

2.11   Diode junctions can also be made with N and P germanium.

2.12   But, silicon and germanium are never mixed when making PN junctions.

2.13   Current will flow through the diode in one direction only.

2.14   The circuit symbol of the diode is as follows:-

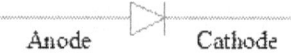

Anode   Cathode

3.1   When the diode is connected in such a way that the current is flowing, it is <u>forward biased</u>.

3.2   In a forward biased diode, the anode is connected to a higher voltage than the cathode, and the current flows.

3.3   Study the way the diode is connected to the battery in the diagram below:-

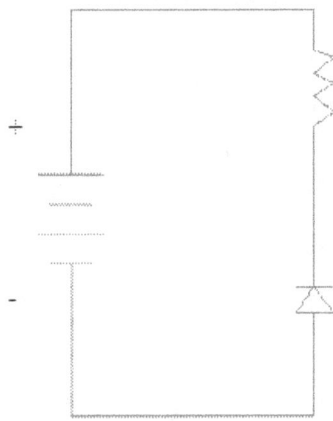

3.3.1 The diode shown above is not forward biased.
3.3.2 The cathode is connected to the higher voltage.
3.3.3 No current can flow as the diode can only conduct current flow in the forward direction.
3.4   When the cathode is connected to a higher voltage level than the anode, the diode cannot conduct.
3.5   In this instance, the diode is said to be <u>back biased</u>, or, <u>reverse biased</u>.
3.6   In many circuits, to simplify calculations, the diode is often considered to be a <u>perfect</u> diode.
3.7   A perfect diode has zero voltage drop in the forward direction and conducts no current in the reverse direction.
3.8   A <u>forward</u> biased perfect diode can be compared to a <u>closed</u> switch.
3.9   It has no voltage drop across its terminals and current flows through it.
3.10   Another similar component is the mechanical switch.
3.11   The mechanical switch has zero voltage drop across its terminals in one condition, and conducts no current in an alternative condition.
3.12   When closed the mechanical switch has no voltage drop across its terminals, and when open it conducts no current.
3.13   A <u>reverse</u> biased diode is similar to an <u>open</u> switch.
3.14   No current flows through it and it will have the full circuit voltage appear across its

terminals.

4.1   The knee voltage for a silicon diode is about 0.7 volt.

4.2   This means that at voltage levels below 0.7 volt, the diode has such a high resistance that it limits the current flow to a very low value.

4.3   This characteristic knee voltage is sometimes referred to as a <u>threshold voltage</u>.

4.4   The knee voltage for a germanium diode is about 0.3 volt.

4.5   The knee voltage is also a limiting voltage.

4.6   In other words, it is the highest voltage that can be obtained across the diode in the forward direction.

4.7   Silicon thus has a higher limiting voltage than germanium.

4.8   At the limiting or knee voltage the diode resistance changes from high to low.

4.9   Knee voltages will be used as the voltage drops across the PN junction when it is forward biased.

4.10 In any given diode the knee voltage will not be exactly 0.7 volt or 0.3 volt.

4.11 It will vary slightly.

4.12 When using diodes in practice - that is, imperfect diodes - two assumptions can be made, which are as follows:-

[1] The voltage drop across the diode is either 0.7 volt or 0.3 volt.
[2] Excessive current is prevented from flowing through the diode by a suitable choice of resistor in series with the diode.

4.13 Actually, all diodes are imperfect.

4.14 The 0.3 or 0.7 voltage values are only approximated.

4.15 Another assumption can be made: The voltage drop across the diode, when it is conducting, is 0 volt.

4.16 This assumes that the ideal characteristic curve indicates low resistance (in this case 0 ohm) immediately upon seeing a small voltage and that the knee is at 0 volt.

4.17 Generally a high resistance (rather than a low resistance) is used to prevent excessive current.

4.18 However, the actual value of the resistance depends on the applied voltage and the maximum current the diode can stand.

4.19 In practice, when the battery voltage is 10 volts or higher, the voltage drop across the diode is often regarded as 0 volt instead of 0.7 volt.

4.20 The assumption here is that the diode is a perfect diode and the knee voltage is at 0 volt rather than a threshold value that must be exceeded.

4.21 This assumption is often used in many electronic design situations.

5.1   Most electronic components have more than 10% tolerance in their nominal values.

5.2   For example, a 1k resistor can be anywhere from 900 ohms to 1,100 ohms and still be valid.

5.3   In practice the value of current through a resistor can be 10% different from that calculated, and can change by 10% if the resistor is changed.

5.4   Calculations are often simplified if the simplification does not change values by over 10%.

5.5   A diode is often considered to be perfect when the circuit voltage is 10 volts or more.

6.1   A current flowing through a diode causes heating and power dissipation, just as with a resistor.

6.2   The following is the power formula for resistors: $P = V \times I$

6.3   This formula can also be applied to diodes to find the power dissipation.

6.4   To calculate the power dissipation in a diode, it is first necessary to calculate the current.

6.5   The voltage drop in this formula is assumed to be 0.7 volt for a silicon diode, even if it is regarded as 0 volt when calculating the current.

6.6   For example, the power of a diode, which has 100 mA flowing through it, is: $P = (0.7) (100 \text{ mA}) = 70 \text{ mW}$

7.1   Diodes are made to dissipate a certain amount of power.

7.2   This is quoted as a maximum power rating in the specifications of the diode by the manufacturer.

7.3   Take a diode with a maximum power rating of 2 watts. How much current can it safely pass?

7.4   The answer is as below:-

$P = V \times I$
$I = P/V$
  $= 2 \text{ watts}/0.7 \text{ volt}$
  $= 2.86 \text{ A}$ (rounded off to two decimal places)

7.5   Provided the current in the circuit does not exceed 2.86 Amperes, the above diode will not overheat and burn out.

Diode Breakdown

8.1   All diodes will break down when connected in the reverse direction if excess voltage is applied to them.

8.2   The breakdown voltage is a function of how the diode is made.

8.3   The breakdown voltage varies from one type of diode to another.

8.4　It is quoted on the specification sheet published by the manufacturer.

8.5　Breakdown is not a catastrophic process.

8.6　In other words, it does not destroy the diode.

8.7　The diode will recover to normal operation if the excessive supply voltage is taken away.

8.8　The diode can be used safely many times again provided the current is limited to prevent it burning out.

8.9　No matter how many times a diode is used it will always break down at the same voltage.

8.10　The breakdown voltage is often termed the peak inverse voltage (PIV) or the peak reverse voltage (PRV).

8.11　The PIVs of some common diodes are as follows:-

Diode	PIV
1N 4001	50 V
1N 4002	100 V
1N 4003	200 V
1N 4004	400 V
1N 4005	600 V
1N 4006	800 V

9.1　Remember: [1] Excessive current can permanently destroy a diode.

　　　　　　[2] Excessive voltage will not harm a diode if the current is limited.

9.2　Burnout is more harmful to a diode than breakdown.

9.3　Breakdown is not necessarily harmful, especially if the current is limited.

The Zener Diode

10.1　Zener diodes are diodes that are manufactured so that breakdown occurs at much lower and more precise voltages than those just discussed.

10.2　Zener diodes are so named because they display the "Zener effect" - a particular form of voltage breakdown.

10.3　A small current will flow through the zener diode at the zener voltage.

10.4　To keep the diode at the zener point this current must be maintained.

10.5　A few milliamperes are all that is required in most cases.

10.6　Shown below are the zener diode symbol and a simple circuit:-

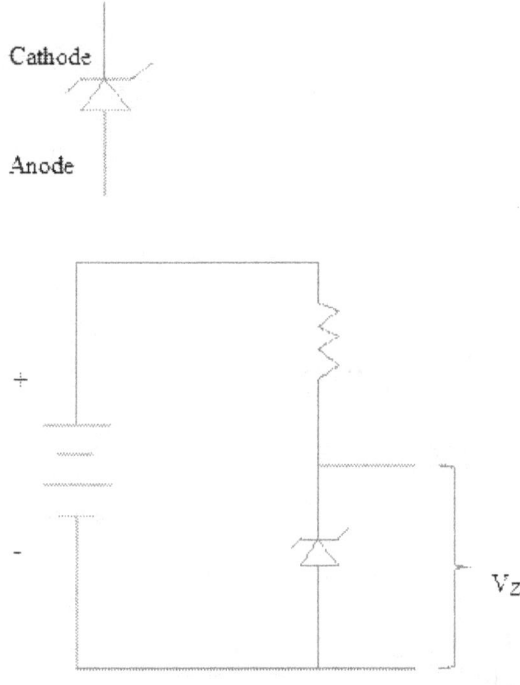

10.7 In the above circuit:

[1] The battery determines the applied voltage.
[2] The zener diode determines the voltage drop, which is labeled V_Z, across it.
[3] The resistor determines the current flow.

10.8 Zeners are utilized to maintain a constant voltage at some point in a circuit.
10.9 Zeners are used for this purpose, while ordinary diodes are not used, because they have a precise breakdown voltage.
11.1 A lowering of voltage across the lamp, or some other component, may be intolerable in many applications.
11.2 This can be prevented by using a zener diode.
11.3 The circuit below illustrates this:-

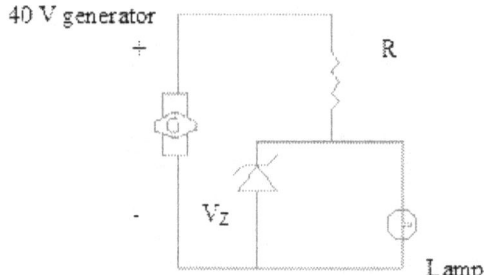

40 V generator

R

V$_Z$

Lamp

[1] In this circuit, a 25 V zener, that is, one which has a 25 V drop across it, is chosen.
[2] The lamp will always have 25 V across it, regardless of what the output voltage is from the generator (provided that the output is always above 25 V).
[3] If the voltage across the lamp is constant, and the generator output drops:

(i) The current through the lamp stays <u>constant</u> because the voltage across the lamp stays constant.
(ii) The current through the zener <u>decreases</u> because the total current decreases.

Review

12.1 Semiconductor diodes are extensively used in modern electronic circuits.

12.2 The following are the main advantages of semiconductor diodes:-

[1] They are very small in size.
[2] They are rugged and reliable if properly utilized. (However, excessive reverse voltage or excessive forward current could damage or destroy the diode.)
[3] Diodes are very easy to use - there are only two connections to make.
[4] They are cheap.
[5] They can be used in all kinds of electronic circuits, from simple DC controls to radio and television circuits.
[6] They can be made in numerous sizes and can handle a wide range of current and power.
[7] Specialized diodes will perform particular functions which cannot be accomplished with any other components.
[8] Diodes are an integral part of transistors.

12.3 All the uses of semiconductor diodes are based on the fact that they conduct in <u>one</u>

direction only.

12.4 Diodes are typically used for the following:-

[1] Logic and decision circuits in computers and control circuits.
[2] Simple switching circuits with no moving parts.
[3] Concerting AC (Alternating Current) to DC (Direct Current).
[4] Recovering the television and radio signals from those transmitted over the air, thereby enabling the person to see and hear the program.

3 ALTERNATING CURRENT (AC)

1.1   Some knowledge of alternating current (AC) is required for electronics.

1.2   An AC is made up of sine waves.

1.3   Sine waves are found in many places and are a perfectly natural phenomenon.

1.4   Probably, the most common is the house current that is provided at a wall plug.

1.5   The house current in the U.S., e.g., is a 120 volt sine wave which changes at a rate of 60 cycles per second (or at a frequency of 60 Hertz). In some countries it is 240 volts.

1.6   A generator, which is a large piece of rotating machinery, in a power station, produces this current.

1.7   In a car, a smaller generator provides the electrical power for the engine and charges the battery.

1.8   The sound produced by most musical instruments consists of sine waves.

1.9   Many other electronic signals, e.g., speech, are complex combinations of many sine waves, all at different frequencies.

1.10   Different voltages cause variations in sine wave currents.

1.11   Circuits, known as oscillators, produce many sine wave voltages and currents.

1.12   Sine waves may change many millions of times a second.

The Generator

2.1   The voltage source is usually a battery, in DC (Direct Current) work.

2.2   The battery produces a steady, constant voltage and a steady, constant current through a conductor.

2.3   The voltage source in AC work is usually a generator.

2.4   The generator produces a regular output waveform, such as the sine wave.

2.5   The following figure shows one cycle of a sine wave:-

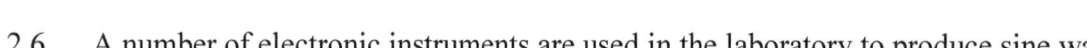

2.6   A number of electronic instruments are used in the laboratory to produce sine waves.

2.7   By turning a dial or pushing a button on the instrument, the voltage and frequency can be adjusted.

2.8   These instruments are known by various names, which are generally based on their method of producing the sine wave or their application as a test instrument.

2.9   The function generator is at present the most popular generator.

2.10   The function generator provides a choice of functions or waveforms, which include the triangle wave and the square wave.

2.11   These waveforms are utilized in testing certain electronic circuits.

2.12   The symbol for a generator is given below (The sine wave that is shown within the circle indicates that it is an AC sine wave source):-

3.1   Look at the diagram of a sine wave below (Certain parameters are marked and the two axes are time and voltage):-

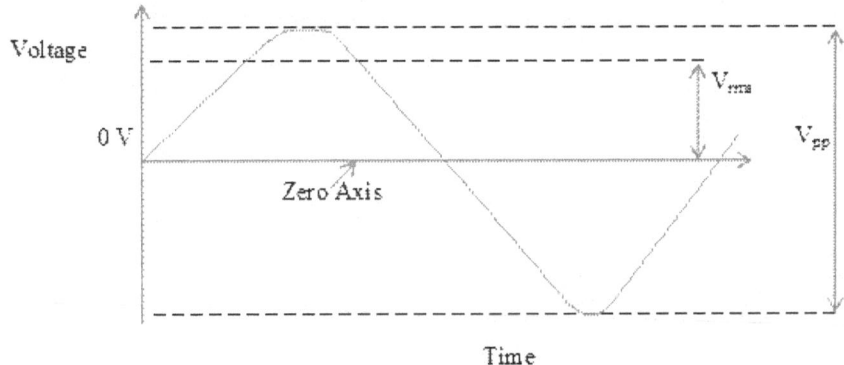

Time

3.1.1   All voltage measurements are made from the zero axis.

3.1.2   The zero axis is also known as the datum line or base line.

3.1.3   Time measurements can be made from any point in the sine wave.

3.1.4   But, time measurements are usually made from the point at which the sine wave crosses the zero axis.

3.1.5   The three most important amplitude or voltage measurements are the:-

[1] peak (p) voltage
[2] peak-to-peak (pp) voltage
[3] root mean square (rms) voltage

3.1.6   The following formulae show the relationship between the p, pp and rms voltages:-

$$V_{pp} = 2 V_p = 2 \times \sqrt{2} \times V_{rms} \qquad [1]$$

$$V_{rms} = 1/\sqrt{2} \times V_{pp}/2 \qquad [2]$$

(Note: $\sqrt{2} = 1.414$, and, $1/\sqrt{2} = 0.707$)

3.1.7   There is one primary time measurement, that is, the duration of the complete sine wave.

3.1.8   All other time measurements are fractions, or sometimes, multiples of this.

4.1   Remember:

[1] One complete sine wave is called a cycle.

[2] The time taken to complete one sine wave is known as the period (T), as shown in the diagram below:-

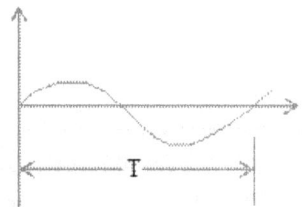

[3] The following formula shows how frequency (f) is related to time (T):-

$$f = 1/T$$

[4] The standard unit for frequency is now Hertz (Hz). An older unit which is sometimes used is cycles per second.

[5] The following could represent electrical AC signals:-

(i) A simple sine wave

(ii) A mixture of many sine waves of different frequencies and amplitudes

Resistors In AC Circuits

5.1   Just as DC currents are, AC signals are passed through components.

5.2   Just as they do with DC currents, resistors interact with AC signals.

5.3   Ohm's law is used to calculate the current that flows through the resistor.

5.3.1 Ohm's law is as follows:-

$$V = RI$$

5.3.2 If an AC signal of 10 V_{PP} is connected across a 10 ohm resistor, the current flow through the resistor is calculated as follows:-

$$I = V/R = 10\ V_{pp}/10\ ohms = 1\ A_{pp}$$

5.3.3 If an AC signal of 10 V_{rms} is connected across a 30 ohm resistor, the current through the resistor is calculated as follows:-

$$I = V/R = 10\ V_{rms}/30\ ohms = 0.33\ A_{rms}$$

(As the voltage is given in rms, the current is also in rms.)

5.4 In the diagram below, 12 V_{pp} is applied to the voltage divider circuit:-

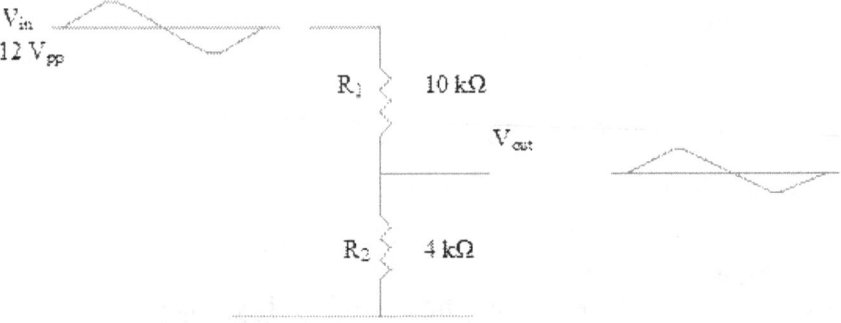

5.4.1 V_{out} is calculated as follows:-

$$V_{out} = V_{in}\ x\ R2/(R1 + R2) = 12\ x\ 4\ k\Omega/(10\ k\Omega + 4\ k\Omega)$$
$$= 12\ x\ 4/14 = 3.43\ V_{pp}$$

Capacitors In AC Circuits
6.1 The capacitor opposes the flow of an AC current.

6.2   The capacitor's opposition to the current flow is called reactance.

6.3   Reactance is similar to resistance in DC circuits.

6.4   The formula for reactance is as follows:-

$$X_c = 1/(2\pi fC)$$

where: X_c = the reactance of the capacitor in ohms
       f = the frequency of the signal in hertz
       C = the value of the capacitor in farads

6.5   Just as with two resistors, a capacitor in series with a resistor will give a voltage divider.

Inductors In AC Circuits

7.1   An inductor is a coil of wire.

7.2   This coil of wire is wound around a piece of soft iron.

7.3   Sometimes, the coil of wire is wound around a non-conducting material.

7.4   The AC reactance of an inductor can be quite high.

7.5   The high AC reactance of an inductor is caused by the magnetic effects of the changing current through the coil.

7.6   The DC resistance is usually quite low.

7.7   The formula for the reactance of an inductor is as follows:-

$$X_L = 2\pi fL, \text{ where } L = \text{value of the inductance in henrys}$$

Resonance

8.1   The inductive reactance <u>increases</u> as the frequency increases.

8.2   The capacitive reactance <u>decreases</u> as the frequency increases.

8.3   If an inductor and a capacitor are connected in series there will be a <u>frequency</u> at which their reactances are equal.

8.4   This frequency is the resonant frequency.

8.5   The resonant frequency is calculated as follows:-

When $X_c = X_L$, $1/(2\pi fC) = 2\pi fL$

Therefore, f (resonant frequency) $= 1/(2\pi \sqrt{LC})$

8.6   Connecting an inductor and a capacitor in parallel will also give a resonant frequency.

8.7 As inductors always have some internal resistance, analyzing a parallel resonant circuit is not as simple as for a series resonant circuit.

8.8 The analysis is similar under certain conditions, however.

8.9 For instance, if the inductor's reactance in ohms exceeds ten times its own internal resistance, the formula for the resonant frequency of an inductor and a capacitor connected in parallel will be the same as if the inductor and the capacitor were connected in series.

8.9.1 This is an approximation that is often used.

8.10 Resonance plays a very important part in electronics.

8.11 Oscillators and filters often rely on resonance.

8.12 Filters are attenuators that will only attenuate a certain band of frequencies and not others.

8.13 Oscillators are electronic circuits that generate a continuous output without an input signal.

8.13.1 Oscillators that use resonant circuits produce pure sine waves.

Comprehensive Review

9.1 In AC circuits, the sine wave is extensively used.

9.2 The function generator is the most common laboratory generator.

9.3 $f = 1/T$

9.4 $I_{pp} = V_{pp}/R$, and, $I_{rms} = V_{rms}/R$

9.5 $V_p = \sqrt{2} \times V_{rms}$, and, $V_{pp} = 2\sqrt{2} \times V_{rms}$

9.6 X_L (inductive reactance) $= 2\pi fL$

9.7 X_c (capacitive reactance) $= 1/(2\pi fC)$

9.8 f (resonant frequency) $= 1/(2\pi\sqrt{LC})$

9.9 Voltage dividers can be formed by capacitors and resistors, and resistors and inductors.

9.10 Voltage dividers and their effects on AC signals play an important role in consumer items, industrial controls and communications.

9.10.1 For instance, such a circuit is utilized to weed out the familiar 60 Hertz hum in audio circuits, and provide the simple "tone controls" on hi-fi equipment.

9.11 Whether an AC signal is a pure sine wave or a complex wave comprising of many sine waves, it is continually changing.

9.12 An AC signal applied on one plate of a capacitor will appear at the other plate.

9.12.1 That is, a capacitor will "pass" an AC signal.

9.12.2 This is illustrated below:-

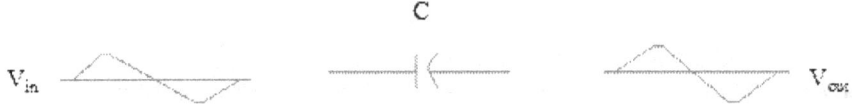

9.13   With an AC signal, a capacitor <u>does not</u> look like an open circuit.

9.13.1 With a DC signal, however, a capacitor <u>does</u> look like an open circuit.

9.13.2 <u>Note</u>: A capacitor is <u>not</u> a short circuit to an AC signal.

9.14   In an AC circuit, a capacitor will pass an AC signal, but will not pass a DC voltage level.

9.15   <u>Remember</u>: A capacitor is neither a short circuit nor an open circuit to an AC signal.

9.16   Generally, a capacitor opposes the flow of an AC current. (This opposition is called the reactance of the capacitor.)

9.17   Reactance is similar to resistance.

9.18   When the input signal frequency changes, the reactance changes.

9.19   If the input is a pure sine wave, the output is also a pure sine wave.

9.19.1 This is illustrated below:-

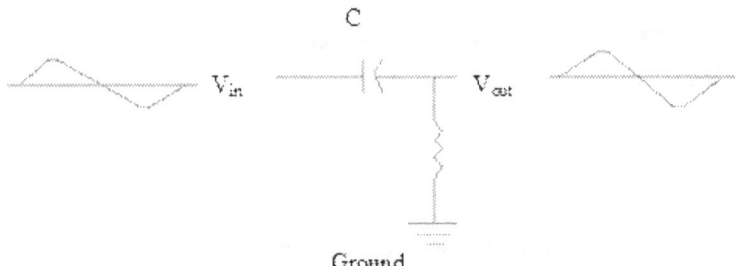

9.19.2 The output sine wave will have the same frequency as the input sine wave.

9.20   A capacitor <u>cannot</u> change the frequency of the input signal.

9.21   With an AC input the capacitor behaves like a resistor. (It is <u>not</u> a short circuit.)

9.22   A capacitor can change the amplitude of the input signal. (In most cases, the output amplitude is different from the input amplitude.)

9.23   Two resistors that are connected can also form a voltage divider.

9.23.1 The formula for the voltage output of the above circuit is as follows:-

$$V_{out} = V_{in} \times R_2/(R_1 + R_2)$$

9.24 Two resistors that are connected in series present a total resistance to the flow of current.

9.24.1 The formula for this total resistance is as follows:-

$$R_T = R_1 + R_2$$

9.25 The opposition to current flow caused by connecting the resistor to the capacitor is called impedance.

9.25.1 The following formula is used to calculate impedance:-

$$Z = \sqrt{X_c^2 + R^2}$$

where: Z = impedance of the circuit in ohms
X_c = reactance of the capacitor in ohms
R = resistance in ohms

9.26 In an AC circuit, the voltage across a capacitor rises and falls at the same frequency as the input.

9.26.1 However, it does not reach its peaks at the same time, nor does it go through zero at the same time.

9.27 The output voltage waveform can either lead or lag the input voltage waveform.

9.27.1 The leading or lagging is measured in degrees found by observing the displacement between the peaks of the two waveforms.

9.28 Note: A half cycle of a sine wave is 180 degrees.

9.29 The displacement between the peaks of the input voltage waveform and the output voltage waveform is known as the phase shift or phase difference.

9.30 The phase shift depends on the value of the frequency, as the value of the impedance and reactance depends on frequency.

9.31 An RC (Resistor/Capacitor) voltage divider with the voltage taken across the capacitor produces a lag in the phase shift of the output voltage.

9.31.1 The figures below are two RC circuits:-

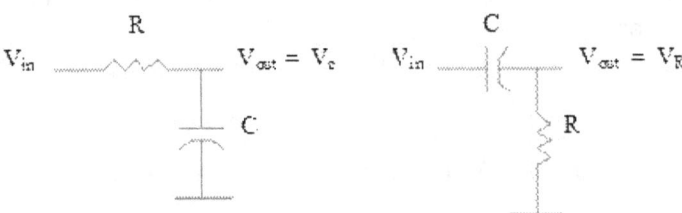

9.32   The figure below is a circuit with a capacitor and a resistor in parallel:-

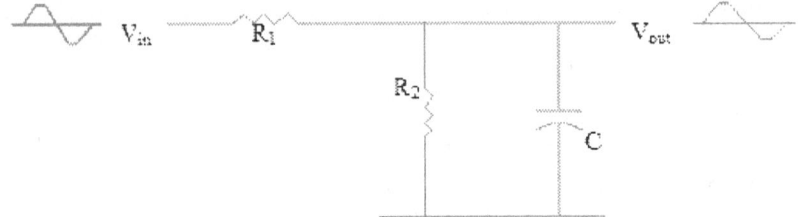

9.32.1 In the above circuit, a DC input signal will not "see" the capacitor. The circuit will work as shown below:

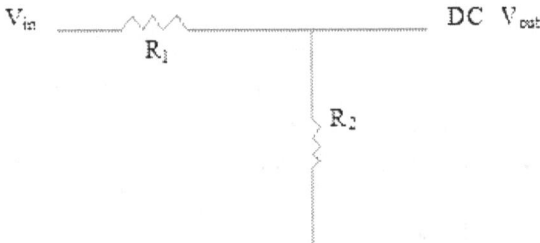

9.32.2 In this same circuit, an AC input will "see" the capacitor. It will see it in parallel with R_2. The circuit will work as shown below, where r is the parallel equivalent of R_2 and X_c:

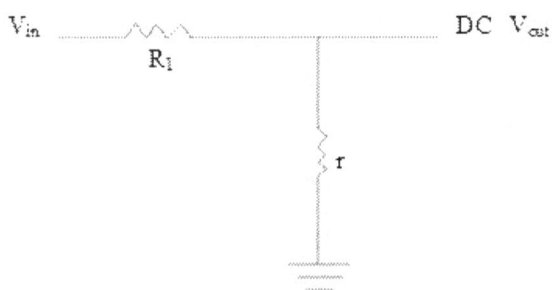

9.33   An inductor can form a voltage divider with a resistor, as shown below:-

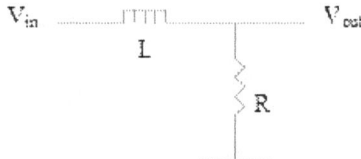

9.33.1 Like the capacitor, the inductor <u>cannot</u> change the frequency of the output voltage.

9.33.2 But, it can change the output voltage's amplitude.

9.33.3 The inductor-resistor series connection above opposes the flow of current.

9.33.4 This current flow opposition is called impedance.

9.33.5 The formula for the reactance of the inductor is as follows:-

$$X_L = 2\pi fL$$

9.33.6 The formula for the opposition to the current flow for the above circuit, or, the impedance of the circuit, is as follows:-

$$Z = \sqrt{X_L^2 + R^2}$$

9.34   In the following voltage divider or circuit, the voltage out (V_{out}) is calculated with the formula shown:-

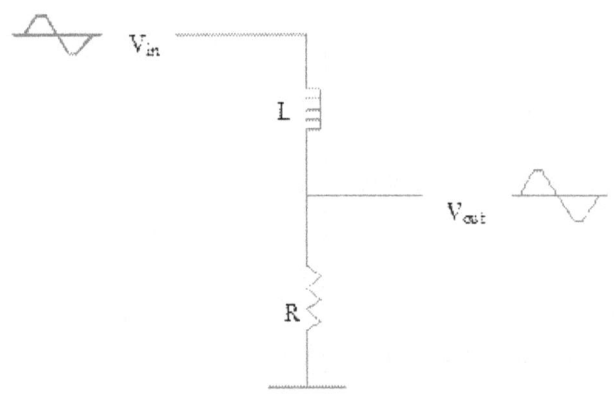

$$V_{out} = V_{in} \times R/Z$$

9.35   The figure below is another voltage divider:-

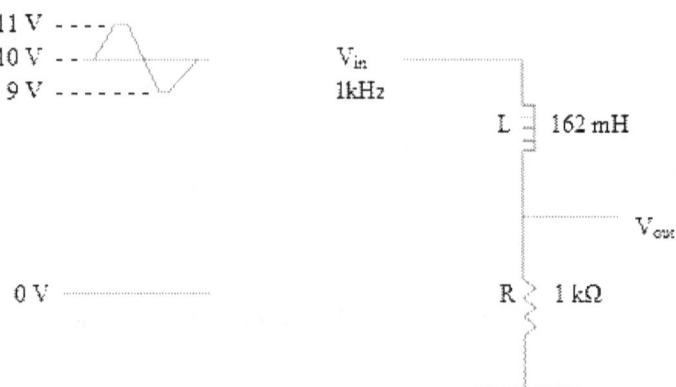

9.35.1 To find the output voltage in the above circuit, the following procedure should be followed:-

[1] Find the DC output voltage, using the DC voltage divider formula. (DC V_{out} = ?)

[2] Find the inductor's reactance. (X_L = ?)
[3] Find the AC impedance. (Z = ?)
[4] Find the AC output voltage. (AC V_{out} = ?)
[5] Combine the outputs to find the actual output. Draw a simple graph.

9.35.2 The above steps yield the following results:-

[1] DC V_{out} = 10 x 1 kΩ/(1 kΩ + 0) = 10 V

[2] X_L = 1 kΩ (approximately)

[3] Z = $\sqrt{1^2 + 1^2}$ = $\sqrt{2}$ = 1.414 kΩ

[4] AC V_{out} = 2 x 1 kΩ/1.414 kΩ = 1.414 V_{pp}

[5]
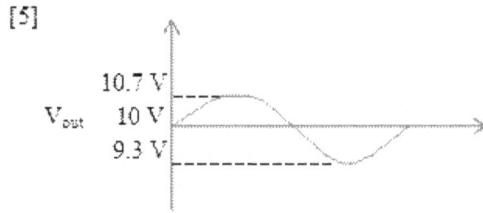

9.36 The inductor will also produce a phase shift in the output signal, as with the capacitor.
9.37 The phase shift or phase difference between the output voltage and the input voltage is measured in degrees.
9.38 Look at the following RL circuits:-

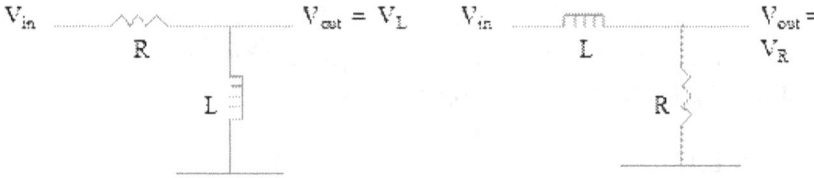

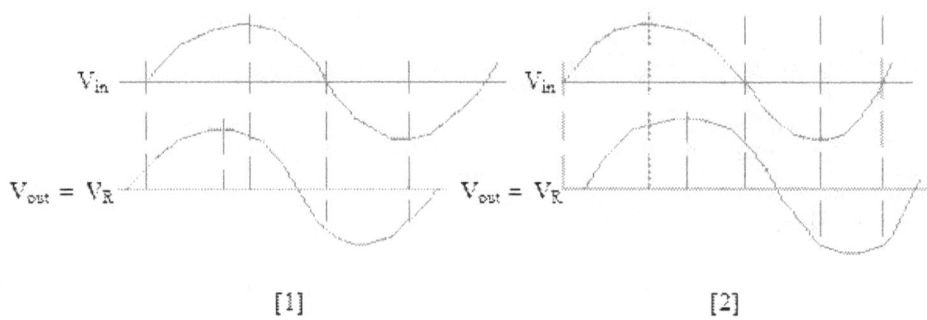

[1]   [2]

9.38.1 In Graph [1], the output voltage <u>leads</u> the input voltage.
9.38.2 In Graph [2], the output voltage <u>lags</u> the input voltage.
9.38.3 The following vector diagram can be drawn for both the above circuits:-

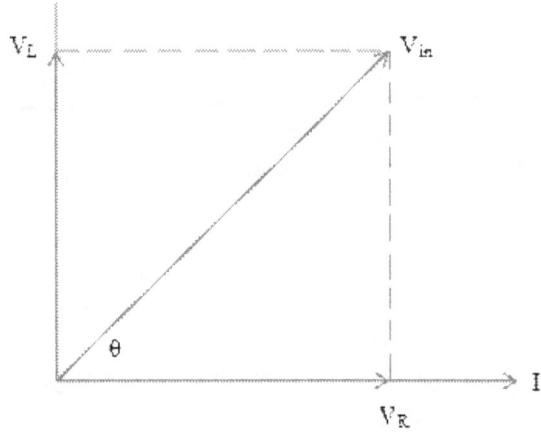

9.38.4 The current through the inductor lags the voltage across the inductor by 90 degrees.
9.38.5 The following formula is utilized to find the phase angle:-

$$\tan \theta = V_L/V_R = X_L/R = 2\pi fL/R$$

4 RESONANT CIRCUITS

1.1   When a capacitor and an inductor are used together in parallel or in series resonance takes place.

Connecting The Capacitor And Inductor In Series

1.2   The figure below shows an inductor and a capacitor connected in series:-

1.3   The above series combination can be utilized to form a voltage divider with a resistor, as shown below:-

1.4   The two circuits shown above are often called RLC circuits, as they contain R (resistance), L (inductance) and C (capacitance).

1.5   Note: The small DC resistance of the inductor can be assumed to be much less than the resistance of the resistor R and can be safely ignored.

1.6   With an AC signal applied to the RLC circuit, both the capacitor and the inductor will have a reactance value which depends on the frequency.

1.7   Remember:

[1] The formula for the capacitor's reactance is:

$$X_C = 1/(2\pi fC)$$

[2] The formula for the inductor's reactance is:

$$X_L = 2\pi fL$$

1.8   The formula for the net reactance of the capacitor and inductor series combination is as follows:-

$$X = X_L - X_C$$

1.8.1 The formula for the impedance of this circuit is as follows:-

$$Z = \sqrt{R_2 + X_2}$$

Connecting The Capacitor And Inductor In Parallel

2.1 The figure below shows the capacitor and the inductor connected in parallel:-

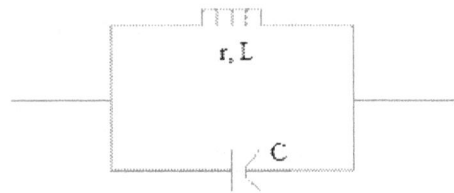

2.1.1 The inductor internal resistance of the above circuit affects the resonance of the circuit and the circuit's operation with regard to frequency response.

2.2 The net reactance is equal to zero in a series circuit.

2.3 The reactances are inverted mathematically and the sum of the inverses is equal to zero, in a parallel circuit.

2.4 The net impedance is equal to the total resistance in the circuit including the coil resistance, in a series circuit which is at resonance.

2.5 The net impedance is more complicated for a parallel circuit.

2.6 The resonance of the parallel circuit is obtained from the following formula:-

$$f_r = 1/(2\pi\sqrt{LC}) \ \ x \ \ \sqrt{1 - (r^2C/L)}$$

2.6.1 If the coil reactance is equal to or more than 10 times the coil resistance, then the formula for the series circuit can be utilized:-

$$f_r = 1/(2\pi\sqrt{LC})$$

2.6.2 In this formula the Q of the coil is equal to or greater than 10, where:

$$Q = X_L/r$$

2.7 Remember:

[1] The formula for calculating the resonant frequency of a parallel circuit if the Q of the coil is equal to or more than 10, say 20, is the same as that for the series circuit, which is as follows:-

$$f_r = 1/(2\pi\sqrt{LC})$$

[2] If Q of the coil is less than 10, say 7, the parallel circuit's resonant frequency is calculated with the following more complicated formula:-

$$f_r = 1/(2\pi\sqrt{LC}) \times \sqrt{1 - (r^2C/L)}$$

(However, if Q is known, the following formula can be used:

$$f_r = 1/(2\pi\sqrt{LC}) \times \sqrt{Q^2/(1 + Q^2)})$$

2.8 The parallel circuit's total opposition (impedance) to current flow can be given by the following formulae at resonance:-

$$Z_p = L/rC \text{ , for any value of Q}$$
$$Z_p = Q^2r \text{ , if Q is equal to or more than 10}$$

2.9 As in the case of the resonant series circuit, at resonance, the parallel circuit's total impedance is considered to be all resistance, as the inductive and capacitive reactance effects have cancelled themselves out.

3.1 The parallel circuit can be used in a voltage divider with a resistor, as shown below:-

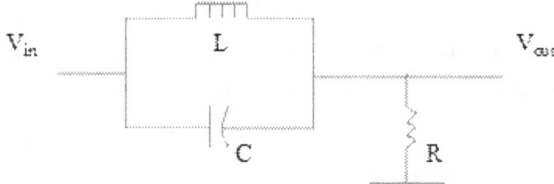

3.2 The total impedance for the voltage divider circuit at resonance is obtained with the following formula:-

$$Z_T = Z_p + R$$

(Note: This is only applicable at resonance, for, at all other frequencies, Z_T is a complicated calculation or formula that is found by considering a series r, L circuit in parallel with a capacitor.)

3.3   At the resonance frequency, the voltage divider circuit's output voltage will be at its lowest value because the parallel resonant circuit's impedance is at its highest value at this frequency.

4.1   Q is utilized to determine how "good" a capacitor-inductor circuit is at rejecting or passing a given frequency or band of frequencies.

4.2   For example, if fr is the main or center frequency, what other frequencies, wanted or unwanted, will be passed or rejected with fr?

4.3   Here, we are talking about the circuit's bandwidth.

4.4   In practice, the bandwidth can be measured or calculated from the component values.

4.5   The following two formulae can be used to calculate the bandwidth:-

$$BW = f_2 - f_1 \qquad [1]$$

$$BW = fr/Q \qquad [2]$$

where: $Q = X_L/R$

4.6   A circuit's Q is a measure of its bandwidth.

Oscillators

5.1   Where frequency rejection or selection is required capacitive and inductive circuits are widely used.

5.2   Capacitive and inductive circuits are also found in oscillators.

5.3   To produce a sine wave output many oscillators use a tuned parallel LC circuit.

5.4   Take a look at the diagrams below:-

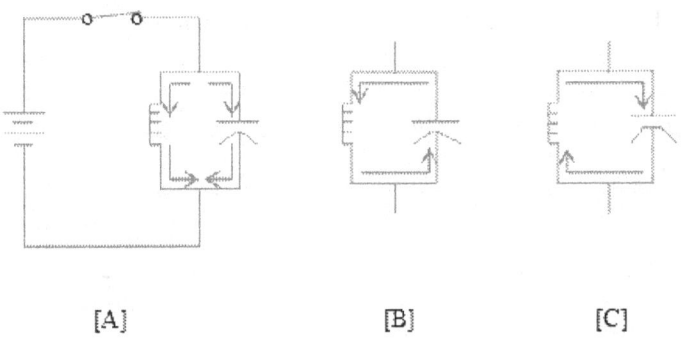

[A]   [B]   [C]

5.4.1   Current flows through the completed circuit when the switch is <u>closed</u>.

5.4.2   It will be hard for the current to flow through the inductor at first, as the inductor tends to oppose any changes in current flow.

5.4.3   It will be easy for the current to flow into the capacitor and cause it to change.

5.4.4   The capacitor passes less current as it charges.

5.4.5   Current now begins to flow through the inductor.

5.4.6   Finally, the capacitor is fully charged.

5.4.7   A steady current now flows through the inductor.

5.5   When the switch is opened, the capacitor discharges through the inductor. (As shown in Diagram [B] above.)

5.5.1   The current maintains the magnetic field as it continues in the same direction through the inductor.

5.5.2   As the capacitor discharges, the current decreases.

5.5.3   No current will flow through the inductor when the capacitor is fully discharged.

5.5.4   The magnetic field collapses as there is no current in the inductor.

5.5.5   The magnetic field induces a current to flow in the inductor as it collapses.

5.5.6   This current flows in the same direction again. (As shown in Diagram [B] above.)

5.5.7   The capacitor is now charged in the reverse direction by this current.

5.5.8   The capacitor now discharges through the inductor again.

5.5.9   The current now flows in the opposite direction. (As shown in Diagram [C] above.)

5.5.10 A magnetic field of the opposite polarity is built.

5.5.11 When the capacitor is fully discharged the magnetic field stops growing.

5.5.12 The magnetic field collapses when there is no current flowing.

5.5.13 This then causes current to flow in the direction shown in Diagram [C].

5.5.14 This current charges the capacitor in the original direction.

5.5.15 The capacitor stops charging when the magnetic field has fully collapsed.

5.6   The capacitor now begins to discharge again.

5.6.1   This causes current to flow through the inductor in the original direction shown in Diagram [B].

5.7   This current flow "see-saw" will go on indefinitely.

5.8   The inductor will have a voltage drop across its ends when the current flows through it.

5.9   As the current varies this voltage varies.

5.10   When viewed on an oscilloscope this voltage would look like a sine wave.

5.11   In a perfect circuit there will be a continuous sine wave.

5.12   In practice, a small amount of power will be lost in the DC resistance of the inductor and the other wiring.

5.13   The sine wave will gradually decrease in amplitude as shown below:-

5.14   After a few cycles, the sine wave will disappear.

5.15   The sine wave can be prevented from fading by replacing a small amount of energy each cycle.

5.16   This lost energy can be passed into the circuit by momentarily closing and opening the switch (shown in Diagram [A] above) at the correct time.

5.16.1 By doing this the oscillations would be sustained <u>indefinitely</u>.

5.17   By having a voltage drop across a few turns of the inductor coil, or, by varying the voltage across the inductor, an electronic switch could be operated.

5.18   Look at the figure below:-

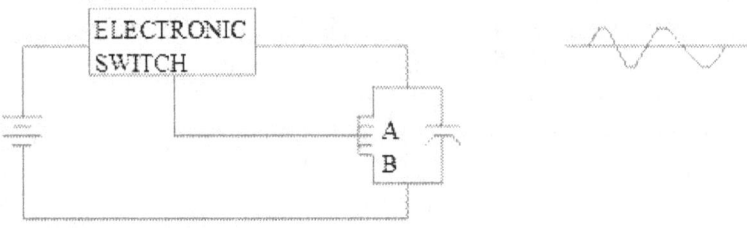

5.18.1 In the above circuit, the small voltage across the few turns between point A, which is about halfway along the coil, and point B, which is at the end of the coil, is utilized to operate the electronic switch.

5.18.2 Using a small part of an output voltage in this way is called "feedback".

5.18.3 What happens is that part of the output voltage is "fed back" to an earlier part of the circuit to get it to operate correctly.

5.19 Such a circuit will produce a continuous sine wave output of constant frequency and constant amplitude when it is properly set up.

5.20 We call this circuit an oscillator.

5.21 The following formula is used to calculate the frequency of the sine waves that are generated by an oscillator:-

$$f = 1/(2\pi\sqrt{LC})$$

5.22 All practical oscillator circuits make use of the same principle, which has been explained above.

5 TRANSFORMERS

1.1   Transformers "transform" an AC voltage to a higher or lower level.

1.2   For instance, in a television set, one transformer produces several thousand volts from a 28 V oscillator, while another transformer produces 28 V from the 120 V power supply.

1.3   Transformers are found in communications and digital equipment, where they are often as small as a finger-nail.

1.4   Transformers can be found near the top of power line poles in the street.

1.5   In factories, transformers can be seen as large fenced off installations.

1.6   In communications equipment, transformers are usually utilized to "isolate" parts to eliminate electrical noise and other static interference.

The Basics Of The Transformer

2.1   Look at the two coils below which are placed very close to one another:-

2.1.1   If an AC voltage is applied to one of the above coils causing a current to flow, a changing (or alternating) magnetic field around the coil will be set up.

2.1.2   As the second coil is very close to this coil, much of the changing magnetic flux cuts the turns of the second coil.

2.1.3   This causes an AC voltage to be induced across the second coil's terminals.

2.1.4   The voltage output of the second coil will be at the <u>same frequency</u> as the input voltage.

2.1.5   Now, the voltage output of the second coil can be used to supply the power requirements of an external circuit.

2.1.6   The way in which the two coils are coupled together (magnetically) and their configuration will vary with different kinds of transformer.

2.2   The transformer is a magnetic circuit device.

2.3   The transformer is a basic component in an electronic circuit.

2.4   The closer the two coils are wound together, the more their inductive coupling will improve.

2.5   The coils are often wound on a common iron core called a "former".

2.6   As magnetic flux exists or flows easily in the iron, this will give maximum coupling.

2.7   Moreover, transformers can have numerous output coils (rather than just two coils).

2.7.1   <u>Remember</u>:

[1] When the two coils of a transformer are wound together, they are <u>not</u> connected

electrically.

[2] The coupling of two or more coils results in a magnetic circuit device.

[3] The transformer is a magnetic circuit device.

[4] If an AC voltage were applied to the terminals of the first coil of a transformer, an AC voltage of the same frequency would appear at the terminals of each of the other coils.

[5] The microphone, the speaker and the relay are some of the other devices which are based on magnetic circuit principles.

3.1   A transformer is an AC (only) device. (A DC voltage which is applied to one set of terminals has no effect on the other terminals.)

3.2   When a sine wave is applied to one coil, a sine wave of the same frequency will appear across the other coil, as shown below:-

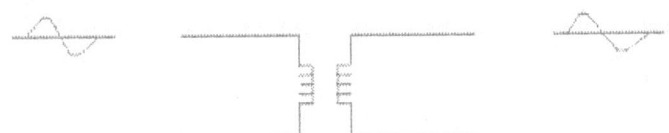

3.3   The primary coil is the coil to which the signal is applied.

3.4   The secondary coil is the other coil in which the induced voltage appears.

3.5   Remember:

[1] The frequency of the signal at the primary coil and the frequency of the signal at the secondary coil will be the same.

[2] If a DC signal is applied to the primary coil, there will be no output at the secondary coil. (DC does not pass through a transformer.)

[3] The DC signal does not produce the changing magnetic field which is required to induce a voltage output at the second coil.

4.1   Look at the diagram below:-

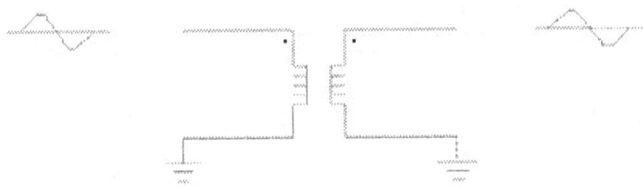

4.1.1   The input and output waveforms can be compared if one side of each coil is grounded, as shown above.

4.1.2   When the input becomes positive, the output also becomes positive. (The output and the input are thus regarded to be <u>in phase</u>, as shown in the above diagram.)

4.1.3   In the above diagram, the ends of the coils which produce the in-phase voltages are indicated by the dots.

4.1.4   The output will be inverted from the input if one coil is reversed. (The output and the input are thus regarded as <u>out of phase</u>.)

4.1.5   The diagram of a transformer with its input and output out of phase is shown below. (Notice the position of the dots now.):-

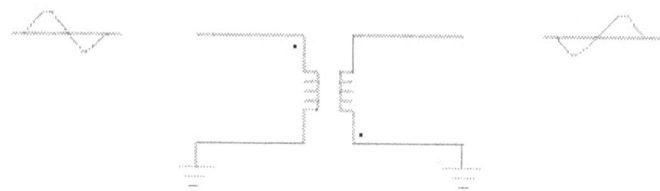

5.1   Look at the diagrams below:-

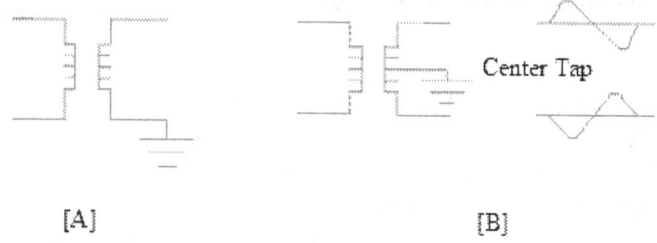

[A]   [B]

5.1.1   One side of the secondary is often (though not always) grounded in a circuit, when there is no center tap in a transformer. (As in Diagram [A].)

5.1.2   In a transformer which has a center tap, the tap is often grounded.

5.1.3   In Diagram [B] above, the output from the two ends of the center-tapped secondary are <u>out of phase</u>.

6.1   The practice and theory regarding transformers are all based on the following one fact:-

Power In = Power Out ($P_{in} = P_{out}$)

6.2 Transformers are imperfect in practice.

6.2.1 A little loss of power normally takes place.

6.3 A perfect transformer is said to have an <u>efficiency</u> of 100%.

6.4 An imperfect transformer's efficiency may be, say, 97%.

6.5 A larger output voltage will be induced in the secondary if the number of turns of wire in the secondary is increased.

6.6 A transformer's output voltage from the secondary is <u>directly proportional</u> to the number of turns in the secondary.

7.1 Look at the diagram below:-

7.1.1 In the above transformer, N_p and N_s represent the number of turns in the primary and secondary coils respectively.

7.1.2 The ratio of the input voltage to the output voltage is similar to the ratio of the number of primary turns to the number of secondary turns.

7.1.3 This is expressed by the following formula:-

$$V_{in}/V_{out} = N_p/N_s$$

(The ratio of the number of primary turns to the number of secondary turns is known as the <u>turns ratio</u> (TR).)

7.2 <u>Note</u>: When V_{in} (and hence V_{out}) is expressed in rms, e.g., 110 Vrms (instead of pp or peak-to-peak), the transformer is sometimes called an isolation transformer, that is used to isolate or separate the load and the voltage source electrically.

7.3 The power supply from the mains (which is 120 V AC in the U.S.A.) has to be stepped down to a more suitable voltage (say 28 V AC) to operate some electronic equipment.

7.3.1 Look at the following figure:-

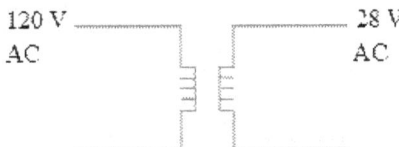

120 V .... 28 V
AC   AC

7.3.2 The turns ratio (TR) for the above transformer is calculated as follows:-

$N_p/N_s = 120/28 = 4.3 : 1$ (approximately)

8.1 Remember the following formulae:-

$V_{in}I_{in} = V_{out}I_{out}$ yields $V_{in}/V_{out} = I_{out}/I_{in} = TR$

8.1.2 The above equation shows that current and voltage are inversely related.

8.1.3 Therefore:-

$I_{out} = I_{in} \times TR$, and,
$V_{out} = V_{in} \div TR$

9.1 A "step up" transformer "steps up" the voltage level, e.g., from 120 V AC to 300 V AC.

9.2 A "step down" transformer "steps down" the voltage level, e.g., from 240 V AC to 28 V AC.

Transformers Found In Communications Circuits

10.1 An input signal is often received by way of a very long interconnecting wire (or line) with an impedance of 600 ohms normally, in communications circuits.

10.2 A telephone line that runs between two cities is a typical example (with the output of the sending equipment also connected to the line).

10.3 When the communications equipment is connected to a load that has the same impedance as the output of the equipment, it works best.

10.4 Communications equipment should have an output impedance of 600 ohms, and should be connected to a 600 ohms line.

10.5 However, most electronic equipment do not have a 600 ohm output impedance.

10.6 Often, a transformer is utilized to connect such equipment to a line.

10.7 For convenience, the transformer is often built into the equipment.

10.8 In this case, the transformer is utilized to "match" the equipment to the line, as shown below:-

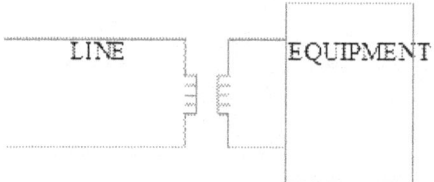

10.9 To exactly match the line, the output of the transformer secondary should have an impedance of 600 ohms.

10.10 The output impedance of the transformer which is measured at the secondary winding of the transformer is governed by the following two things:-

[1] The equipment output impedance.
[2] The transformer's turns ratio. (Note that the wire's DC resistance has no effect and can be ignored.)

11.1 Look at the figure below:-

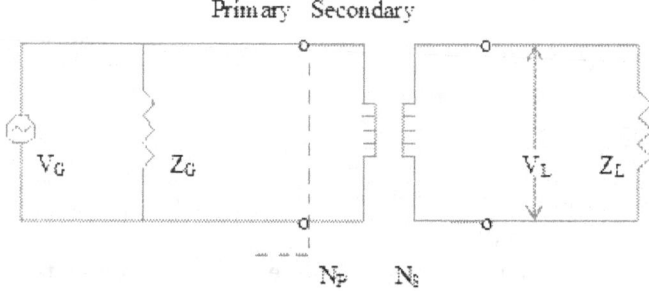

11.1.1 In the above figure, the signal generator which has an output impedance of Z_G is linked to one side of the transformer.
11.1.2 The impedance is in parallel.
11.1.3 A local impedance of Z_L is connected to the other side of the transformer.
11.1.4 The following equation for power can be used:-

Power $= V^2/Z$

11.1.5 Look at the following equations:-

$$P_{in} = P_{out}$$

Therefore:

$$V_G^2/Z_G = V_L^2/Z_L$$
$$Z_G/Z_L = (V_G/V_L)^2$$

As V_G and V_L are directly across the transformer primary and secondary, therefore:

$$Z_G/Z_L = (N_P/N_S)^2 = (TR)^2$$

11.1.6 It is now obvious that the ratio of the impedances on each side of the transformer is equal to the square of the turns ratio.

11.1.7 Using this relationship, the transformer can be designed to "match" the output impedance of the generator (Z_G) with the load impedance (Z_L).

11.1.8 In other words, the transformer will be designed with a particular turns ratio to "match" the two impedances, ZG and ZL (such that ZG will equal ZL).

11.1.9 Note: [1] Matching will work at all frequencies.

[2] It is not necessary to take the frequency of the generator into consideration when performing the above calculations.

[3] But, there is a frequency range for all transformers.

[4] Therefore, it is necessary to use a transformer at a frequency which is within the range it is designed for.

Review

12.1 Transformers are made with a given turns ratio (TR).

12.2 The impedances between which the transformers have to be connected are often stated, e.g., "Pri: 12 kΩ, Sec: 4 kΩ" is marked on a transformer.

12.3 The turns ratio will not be stated on a transformer.

12.4 Transformers come in many different sizes.

12.5 A transformer is usually designed to work over a given frequency range. (A transformer which is designed for audio work, e.g., is not suitable for use in transforming the 120 V AC power supply from the mains to a lower voltage.)

12.6 Transformers can be put to the following important uses:-

[1] Step down the 120 V AC power supply from the mains to a lower AV voltage (which is then converted to a DC voltage in a power supply circuit).

[2] AC communication signals, e.g., music, television and voice, from a faraway source can be connected through a transformer to the input of electronic equipment, such as in the cases of telephones and broadcasting.

[3] The output of communications equipment can be connected through a transformer to the lines between cities so that messages may be sent across the lines.

[4] Transformers are often utilized to couple together the various stages in amplifiers, e.g., the transformer is often used to couple the output of the audio amplifier to the speaker.

12.7 The transformer can change impedances and voltage levels of signals. (Maximum voltage or maximum power can be transferred from one equipment to another.)

6 AC DIODE CIRCUITS AND POWER SUPPLY CIRCUITS

1.1   A power supply circuit is found in most electronic equipment.

1.2   The power supply circuit takes the AC supply from the mains and converts it to a DC voltage which is used to provide power for the electronic circuits.

1.3   Power supply circuits are simple in principle.

1.4   Diodes are an important component in power supply circuits.

1.5   How diodes behave in the presence of an AC signal is the basis of how a power supply circuit works.

Diodes In AC Circuits

2.1   Diodes are used for different objectives in AC circuits, where their ability to conduct in only one direction is required.

2.2   Look at the following circuit:-

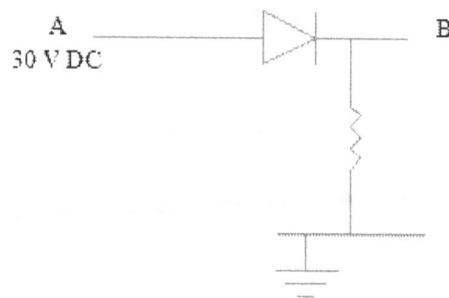

2.2.1   In the above circuit, where 30 V DC is applied at point A, the output voltage at point B will be 30 V DC. (The voltage drop of 0.7 V across the diode is ignored for now.)

2.2.2   If a 15 V DC is applied at point B, the voltage at point A will be 0 V.

3.1   The following circuit has a 20 V_{pp} AC signal which is superimposed on the 20 V DC level:-

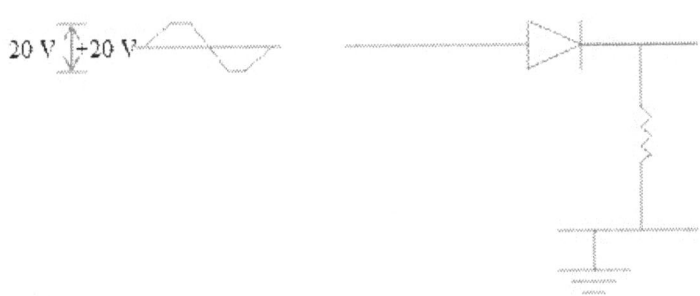

3.1.1 In the above circuit, the positive and negative peak voltages of the input are 30 V and 10 V respectively.

3.1.2 As the diode is always forward biased and hence it always conducts, the output of the above circuit is exactly the same as the input.

4.1 The following circuit has a 30 V_{pp} AC signal which is centered around 0 V DC, or ground:-

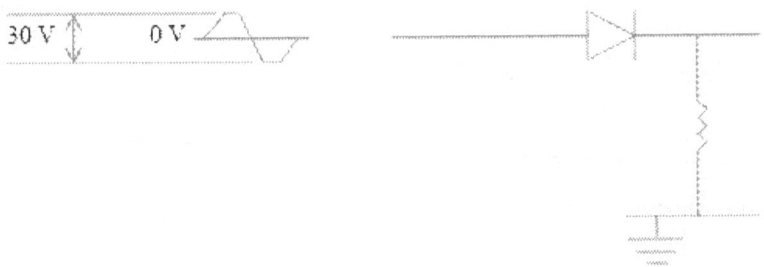

4.1.1 In the above circuit, the positive and negative peak voltages of the input signal are +15 V and -15 V respectively.

4.1.2 In the above circuit, for the positive half wave of the input, the output is +15 V.

5.1 The diode is not <u>non-conducting</u> when the input is <u>negative</u>.

5.1.1 When the input is negative, the output voltage stays at 0 V.

5.2 An input waveform is shown below:-

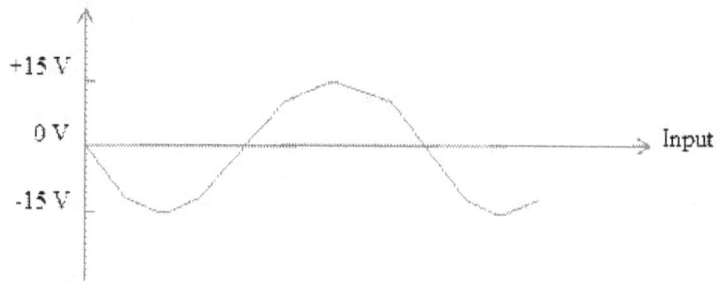

5.2.1 The output waveform for this input is shown below:-

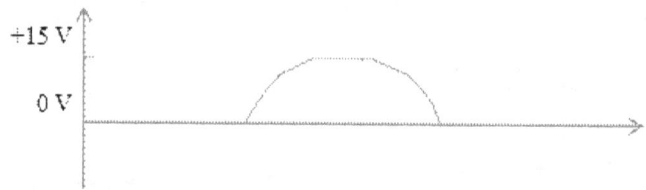

5.2.2 The above is the output waveform for <u>one complete cycle</u> of the input waveform.

6.1 A reversed diode conducts on the negative parts of the cycle and will be reverse biased on the positive parts of the cycle.

6.1.1 This inverts the output waveform.

7.1 Below is the output waveform for three complete cycles of the input waveform of a forward biased diode:-

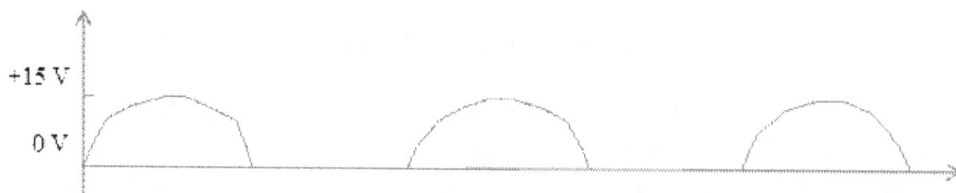

7.1.1 If a reversed diode has been used instead, the output waveform would appear as follows:-

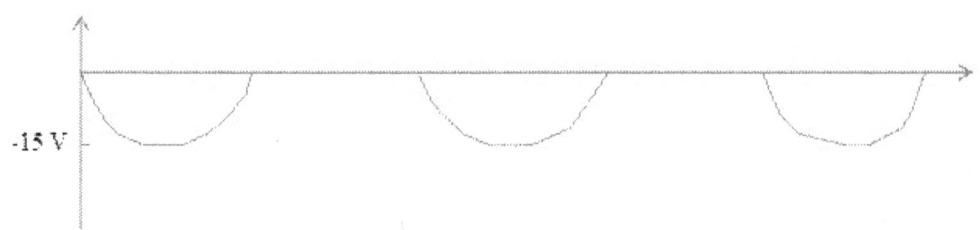

8.1   <u>Remember</u>: As long as the input signal remains negative, the diode will not be forward biased.

9.1   <u>An example</u>:
A circuit with a 20 V_{pp} AC input signal superimposed on a -20 V DC level will have an output voltage of 0 V.

10.1   Diodes are all the time used in electronic power supply circuits.

10.2   The figure below shows a diode connected to the secondary coil of a transformer:-

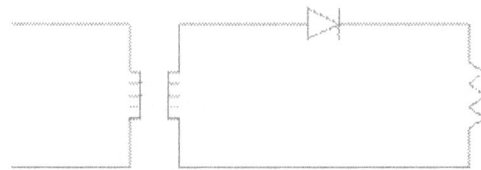

10.2.1 The diode in the above circuit will <u>rectify</u> an AC circuit.

10.2.2 If the secondary of the above transformer is a 40 V_{pp} AC signal centered around 0 V, the waveform of the voltage across the load would appear as follows:-

(This is called <u>half-wave</u> rectification.)

10.2.3 If the diode is reversed, there will be <u>negative half waves</u> (which can be seen on the oscilloscope), that are as follows:-

11.1   Look at the following diagrams:-

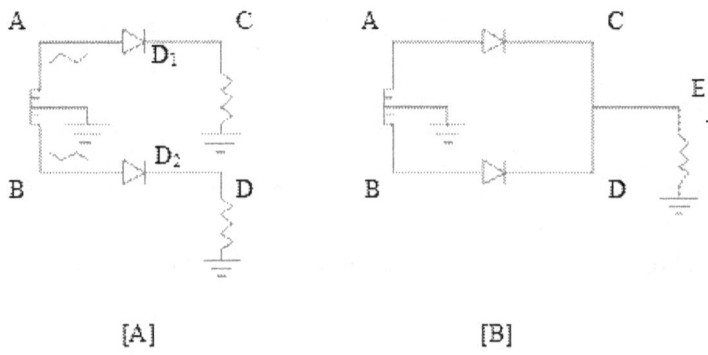

[A]   [B]

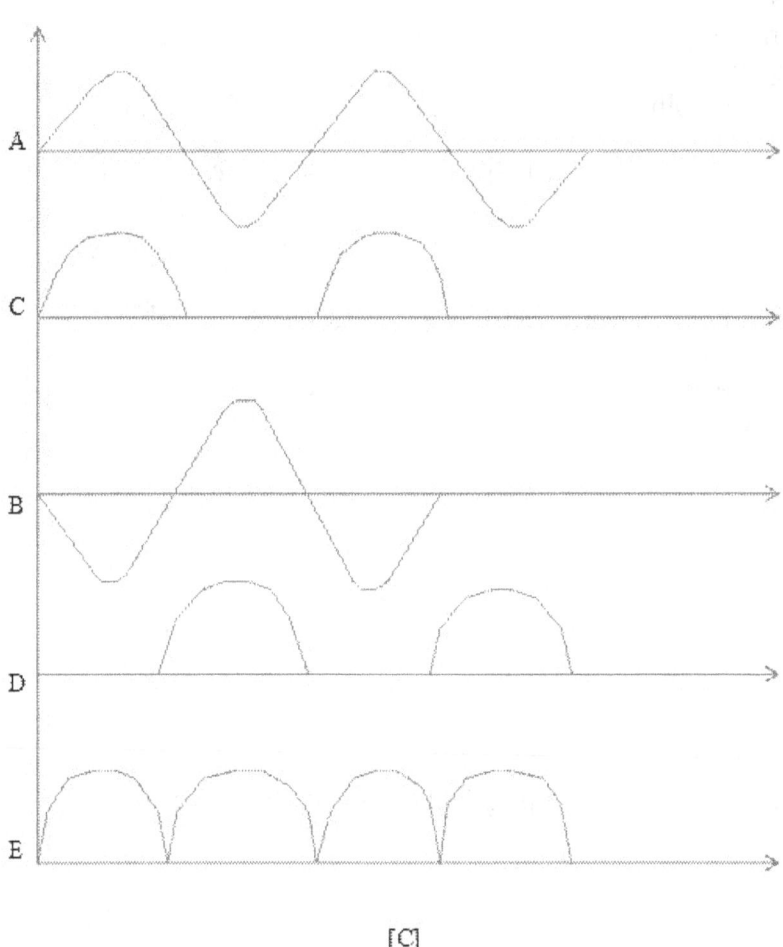

[C]

11.1.1 In Diagram [A], the rectifier circuit's transformer secondary has a center tap.

11.1.2 On the first half wave, diode D_1 conducts and diode D_2 is reverse biased.

11.1.3 On the second half wave, diode D_2 conducts and diode D_1 is reverse biased.

11.1.4 For Diagram [A], the input waveforms from point A and point B to D_1 and D_2 respectively, and the output waveforms at point C and point D, will appear as those shown at points A, B, C and D in Diagram [C].

11.1.5 Diagram [B] is a modification of Diagram [A], where the two diodes are connected across the same resistor.

11.1.6 For Diagram [B], the combined output waveform at point E will appear as that shown shown at point E in Diagram [C].

11.1.7 Note: [1] The output waveforms at point C and point D in Diagram [C] are the result of half-wave rectification.

[2] The output waveform at point E in Diagram [C] is the result of full-wave rectification.

[3] For Diagram [A], the frequency of each of the two rectified half-wave output is 60 Hz.

[4] For Diagram [B], the frequency of the rectified combined output, which is a full waveform, is 120 Hz.

12.1 Compared to half-wave rectification, full-wave rectification of AC gives a much "smoother" conversion of the AC to DC.

12.2 The full-wave rectifier can also be constructed by using a transformer without a center-tapped secondary, as shown below:-

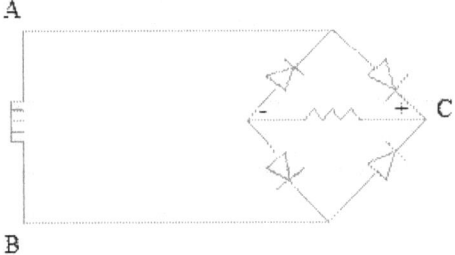

12.2.1 When point A becomes positive, the conduction path will be as shown below:-

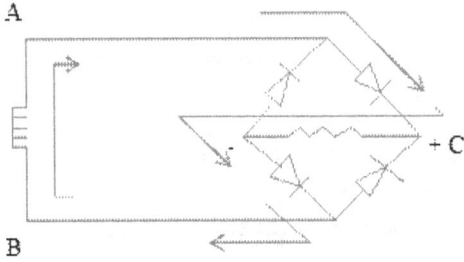

12.2.2 When point B becomes positive, the conduction path will be as shown below:-

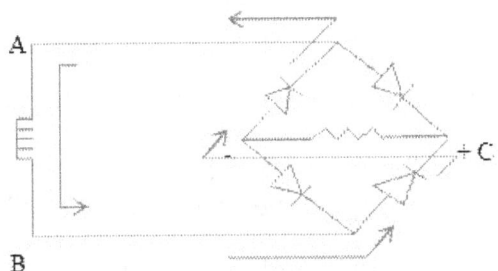

12.2.3 <u>Note</u>: [1] In both the above cases, the direction of current through the load is <u>similar</u>.

         [2] In each conduction path (in each case), the current travels through <u>two</u> diodes.

         [3] The voltage waveform at point C in each case is as shown below:-

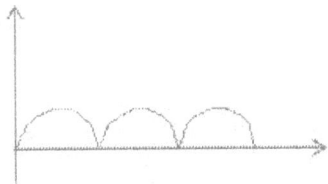

13.1 Often, a rectified AC is regarded as a "pulsating DC".

13.2 The rectified AC or the pulsating DC is most often found in power supply circuits.

Power Supply Circuits

14.1 Basically, the power supply circuit can be divided into four sections: the transformer, the rectifier, the smoothing section, and the load, as shown below:-

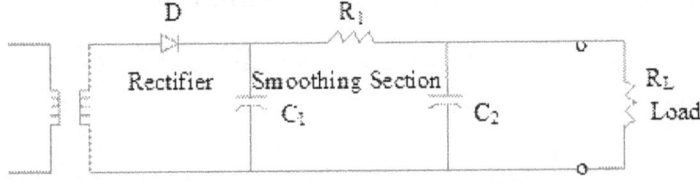

14.1.1 In the power supply circuit (as shown above), a transformer is utilized to convert the 120 V AC (or 240 V AC from the mains in some countries) to an AC voltage which is lower (the choice of this AC level depending on the final level of the DC that is required).

14.1.2 In the above power supply circuit, the output of the rectifier section will be a <u>rectified AC</u> or <u>pulsating DC</u>.

14.1.3 In the above power supply circuit, <u>half-wave rectification</u> will result.

14.2   <u>Remember</u>: If a center-tapped transformer is utilized, <u>two</u> diodes will produce a full-wave output.

14.3   <u>Note</u>: [1]   The transformer in the above power supply circuit has a secondary which has no center tap.

        [2]   In the above power supply circuit, a single diode half-wave rectifier is used.

        [3]   In the above power supply circuit, the smoothing section consists of a resistor (R_1) and two capacitors (C_1 and C_2).

        [4]   The output from the rectifier section will be a half-wave pulsating DC, which is shown below:-

        [5]   This output from the rectifier section (which is a half-wave pulsating DC) is applied to the smoothing section and the load.

        [6]   As this half-wave pulsating DC rises to its peak value, it charges capacitor C_1 to this peak value.

        [7]   When the input drops back to 0 V, the diode becomes reverse biased.

        [8]   When the input drops back to 0 V and the diode becomes reverse biased, capacitor C_1 <u>cannot discharge</u> through the diode.

        [9]   Capacitor C_1 can only discharge through resistor R_1 and the load R_L.

        [10] The value of the capacitor C_1 should be such that the <u>discharge time constant</u> is <u>long</u>.

        [11] If, at the input, there are no further half waves, the waveform will be as follows:-

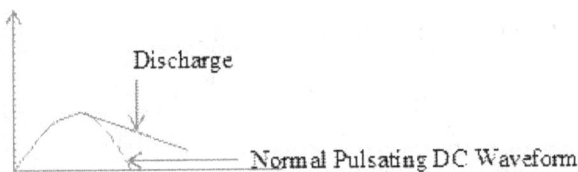

[12] If another half wave arrives, the input and output waveforms will appear as follows:-

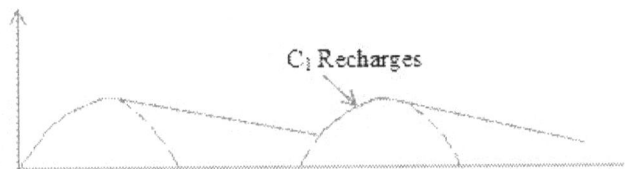

[13] Capacitor C_1 only discharges a small amount before it is recharged to peak value.

[14] This same recharging effect will be repeated when further half waves arrive.

[15] When capacitor C_1 repeatedly discharges, the resulting waveform at capacitor C_1 will be as follows:-

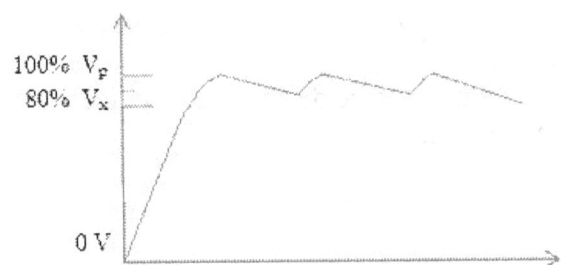

[16] The above waveform is actually a DC level that varies between V_p and V_x.

[17] V_x will be about 80% if the discharge time constant of capacitor C_1 through resistor R_1 and load R_L is made about 10 times the duration of the input pulses.

[18] The "smoothing" will be <u>better</u> if the discharge time selected is <u>more than 10 times</u> the duration of the input pulses.

[19] The average DC output level of the above circuit can be estimated to be about 90% of V_p.

14.4   <u>Also Note</u>: [1] The smoothing section takes the pulsating DC and converts it to a "pure" DC with as little AC "ripple" as possible.

[2] The smoothed DC is next applied to the load.

[3] Study the two figures shown below:-

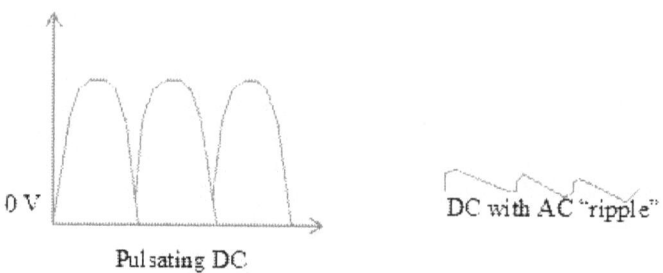

Pulsating DC

DC with AC "ripple"

[4] The load is driven by the power supply.

[5] The load can be a complex electronic circuit or a simple lamp.

[6] The load draws a certain current and requires a certain voltage across its terminals.

[7] Therefore, the load will have a resistance.

[8] The current and voltage required by the load, and hence the resistance, are usually known.

[9] The power supply circuit has to be designed to provide the required current and voltage to the load.

14.5   <u>Remember</u>:

[1] The smoothing section of a power supply circuit converts the pulsating DC to a "pure" DC.

[2] A load such as an electronic circuit or a lamp is connected to a power supply.

[3] In most cases, for the purpose of power supply circuit design, the load can be treated like a resistor.

7 TRANSISTORS

1.1   Without doubt the transistor is the most important modern electronic component.

1.2   It was discovered in 1948.

1.3   It has made great and profound changes in electronics and in our daily lives.

1.4   The transistor is an electronic component that acts similarly to a simple mechanical switch.

1.5   It is actually used as a switch in much modern electronic equipment.

1.6   Like a mechanical switch, it can be made to conduct or not conduct an electric current.

2.1   A transistor can also be made to operate as an amplifier.

2.2   The transistor-amplifier reproduces an output that is a magnified version of an input signal.

2.3   Besides using the transistor to change from one state to another (ON or OFF), we can use it to increase voltage, current, or power levels.

2.4   In many electronic circuits, amplification is required.

2.5   Transistors are found in computers and other modern circuits.

3.1   The bipolar junction transistor, commonly called the BJT, is the most common transistor type.

3.2   It is not the only transistor type.

3.3   Another type of transistor is the junction field effect transistor, or JFET.

3.4   The BJT is the most common transistor.

4.1   There are different packaging techniques for several common transistor case designs in use today.

4.2   Transistors are designed to be either soldered directly into a circuit or inserted into sockets which are designed especially for them.

4.3   When soldering a transistor great care has to be taken, as the transistor can be damaged if either the leads or case are overheated.

4.4   Whenever soldering to transistor leads, proper heat sinking should be used.

4.5   Connections may be soldered to socket terminals before a transistor is inserted to offer complete protection from heat sources.

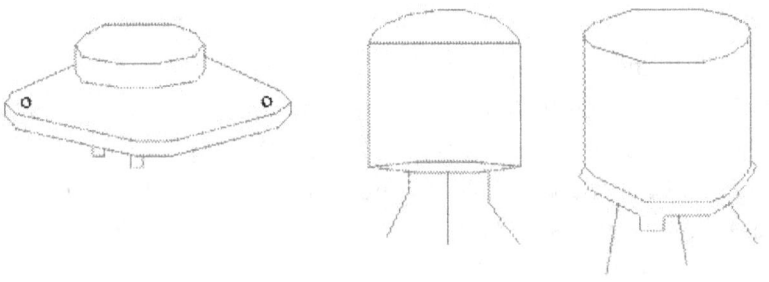

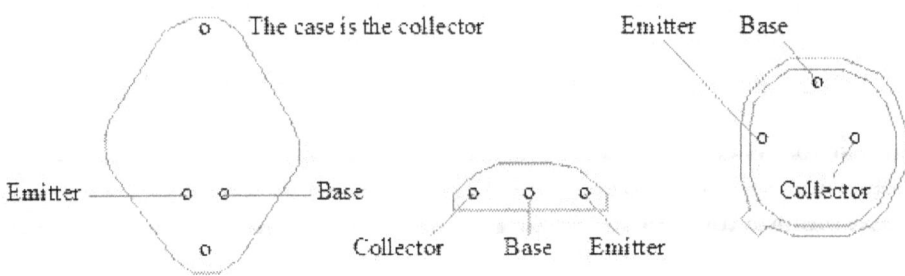

4.6   There are three leads on most transistors.

4.7   Where there are only two leads, the case takes the place of the third lead. (This applies only to power transistors.)

4.8   The three leads or connections of a transistor are called emitter, base and collector.

4.9   <u>Remember</u>: Great care should be exercised when soldering transistors into a circuit as excessive heat can damage the transistors.

5.1   A transistor can be considered as two diodes connected back to back, in its simplest form.

5.2   This is shown in the drawings below:-

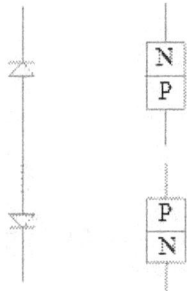

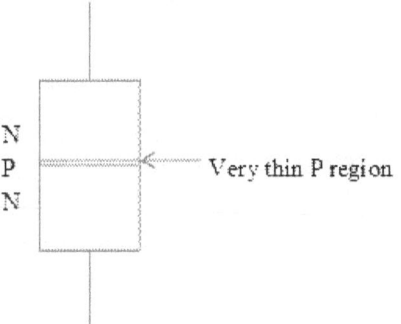

5.3   One very important modification is made in the construction process - instead of only two separate P regions as shown, only one very thin region is used:

5.4   Two separate diodes joined back to back will behave like a transistor, and not like two diodes.

5.5   This is accountable by semiconductor physics and not electronics.

5.6   The main construction difference between two diodes connected back to back and a transistor is the very thin P region used in the transistor.

5.7   The connections of the three terminals of a transistor, the base, the emitter and the collector, are shown below:-

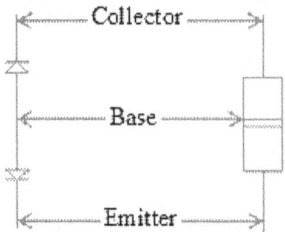

5.8   Usually, the two diodes are called the <u>base-emitter diode</u> and <u>the base-collector-diode</u>.

6.1   The figures below show transistors with an NPN configuration:-

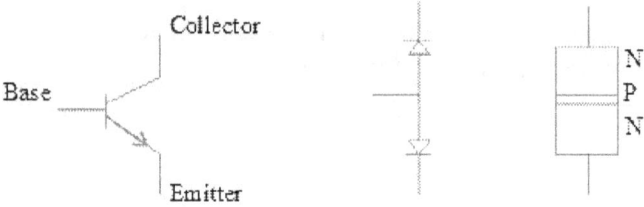

6.2   Transistors with a PNP configuration are shown below:-

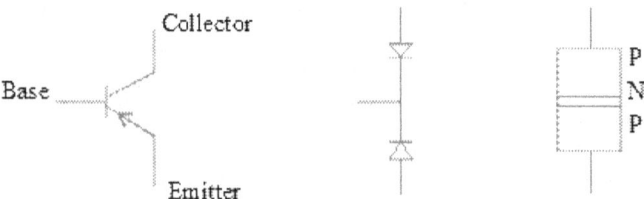

6.3   Both types of transistor, NPN and PNP, are made from either silicon or germanium.

6.4   Silicon and germanium are <u>never</u> mixed in any semiconductor.

6.5   Consider the following example:

    If a battery is connected to an NPN transistor, as shown below, a current will flow as shown.

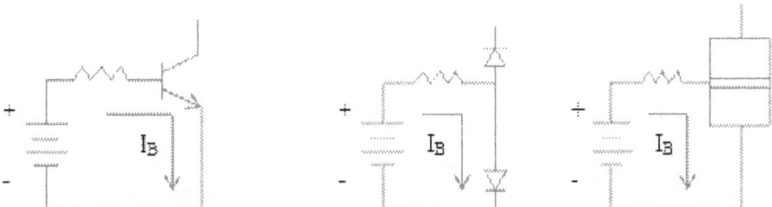

6.5.1   The current that flows through the base-emitter diode is called base current and is denoted by the symbol I_B.

6.5.2   If the battery were reversed base current would not flow as the diode would be back biased.

7.1   Look at the following circuit:-

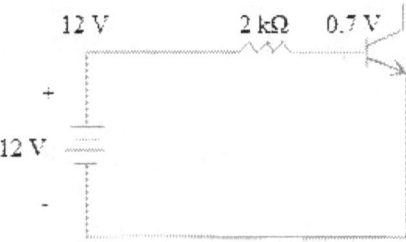

7.1.1   In this circuit since the 12 V battery is much higher than the 0.7 V diode drop, the base-emitter diode can be considered a perfect diode.

7.1.2   Thus the voltage drop can be assumed as 0 V.

8.1   Look at the figure below:-

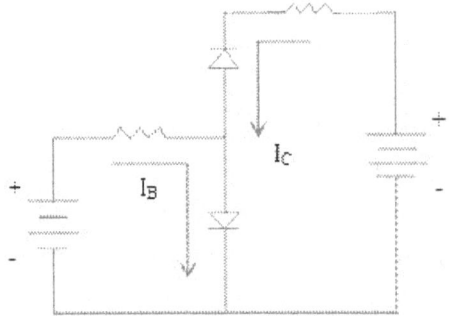

8.1.1   Two circuits are put together.

8.1.2   There are two batteries, as shown above, one in each of the base and collector circuits.

8.1.3   When both the base and the collector circuits are connected, it shows the outstanding characteristic of the transistor.

8.1.4   This is sometimes called <u>transistor action</u>: Collector current will also flow if base current flows in a transistor.

8.1.5   The current that flows through the base-collector diode is the collector current - I_C.

8.1.6   Both the base current (I_B) and the collector current (I_C) flow through the base-emitter diode.

8.1.7   The base current causes the collector current to flow.

9.1   Apart from the NPN transistor, the PNP transistor could have been used.

9.2   There is no difference in how the NPN and the PNP transistors work or behave.

9.3   What is true for one is equally true for the other.

9.4   There is one important circuit difference which is shown below:-

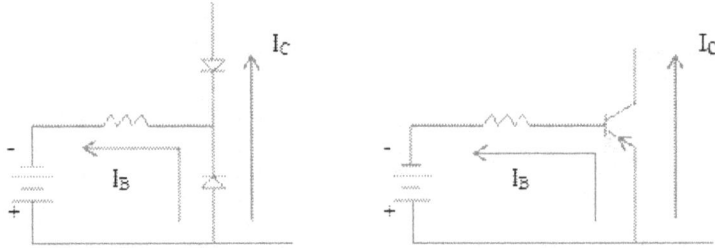

9.4.1   This difference is caused by the fact that the PNP is made with the diodes in the reverse direction from the NPN.

10.1 As mentioned earlier there is absolutely no difference between the NPN and the PNP transistors.

10.2 Both are used equally in electronic circuits.

10.3 No one is favored more than the other.

10.4 The base current causes the collector current to flow in both.

11.1 The ratio of collector current to base current is always <u>constant</u>.

11.2 The collector current is always much bigger than the base current.

11.3 The constant ratio of the two currents is known as the <u>current gain</u> of the transistor.

11.4 This ratio or current gain is a number that is much larger than 1.

11.5 The current gain is denoted by the symbol β, called beta.

11.6 The typical values of β range from 10 to 300.

11.7 A good typical value for many transistors is 100.

11.8 <u>Remember</u>: The collector current is always larger than the base current.

12.1 <u>Note</u>: [1] β is referred to in manufacturer's specification sheets as h_{FE}.
   [2] It is technically referred to as the static or DC_β.
   [3] The h symbol refers to an h parameter.
   [4] There are many h parameters or transistor parameters in general.

12.2 The mathematical formula for current gain, β, is shown below:-

$$\beta = I_C/I_B$$

where:

I_B = Base Current
I_C = Collector Current

12.3 If there is no base current, the collector current will <u>not</u> flow.

12.4 If there is more base current, more collector current will flow.

12.5 <u>Remember</u>: The base current controls the collector current.

13.1 The current gain is a physical property of the transistor.

13.2 The value of the current gain can be obtained from the manufacturer's published data sheets or it can be determined experimentally by the technician or engineer.

13.3 In general, the value of β varies from one transistor type to the other.

13.4 However, it remains constant for a given transistor.

13.5 Transistors of the similar kind have β values which are within a narrow range of each other.

13.6   One of the most often carried out calculations in transistor work is determining the values of either the collector or base current, when β and the other current are given.

The Junction Field Effect Transistor (JFET)

14.1   The JFET, like the BJT, is used in many amplification and switching applications.

14.2   When a high input impedance circuit is needed, the JFET is preferred.

14.3   Compared to the JFET, the BJT has a relatively low input impedance.

14.4   The JFET has three terminals, like the BJT.

14.5   The terminals of the JFET are known as the source, drain and gate and are similar in function to the emitter, collector and base, respectively, of the BJT.

15.1   The semiconductor material of the JFET has a channel which is made of the opposite semiconductor material running through it.

15.2   The JFET is called an N-channel JFET if the channel is of N material.

15.3   The JFET is called a P-channel JFET if the channel is of P material.

15.4   By controlling the effective width of the channel (allowing more or less current to flow), the gate controls the current flow through the drain and the source.

15.5   Just as the voltage on the base of a BJT acts to control the collector current, the voltage at the gate acts to control the drain current.

15.6   The figures below show the N-channel and P-channel JFETs:-

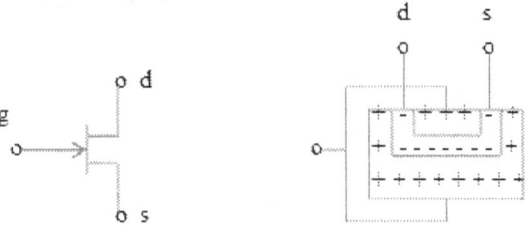

N-channel JFET

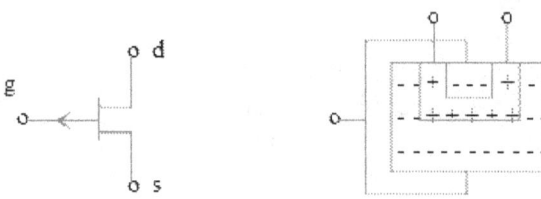

P-channel JFET

15.7   N material uses electrons as the majority carrier.

15.8   The N-channel JFET uses electrons as the primary charge carrier for the drain current.

15.9   Changing the voltage at the gate of the JFET changes the current in the drain; the channel width is controlled electrically by the gate potential.

16.1   A positive voltage is applied to the drain with respect to the source in order to operate the N-channel JFET.

16.2   This enables a current to flow through the channel.

16.3   The JFET will be in the ON condition and the drain current will be at its largest value for safe operation, if its gate is at 0 V.

16.4   A negative voltage applied to the gate will result in the drain current being reduced.

16.5   As the gate voltage becomes more negative, the drain current lessens until cut-off.

16.6   When the drain current is cut-off, the JFET is in the OFF position.

16.7   The figure below shows a typical biasing circuit for the N-channel JFET:-

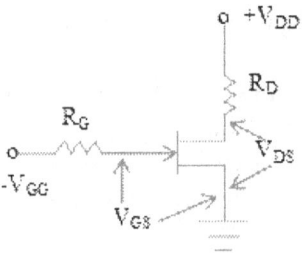

16.8   The polarity of the bias supplies is reversed, for a P-channel JFET.

16.9   The JFET is turned ON with 0 V at the gate, whereas the BJT is turned ON with a large voltage at the gate.

16.10   The JFET is turned OFF with a large current at the gate, whereas the BJT is turned OFF with 0 V at the base.

16.11   The JFET is a "normally ON" device.

16.12   The BJT is a "normally OFF" device.

16.13   The JFET can be used as a switching device.

16.14   Look at the figure of the biasing circuit for the N-channel JFET above.

16.14.1 When the gate to source voltage is 0 V ($V_{GS} = 0$), the drain current will be at its maximum or saturation value.

16.14.2 When the gate to source voltage is 0 V ($V_{GS} = 0$), the N-channel resistance will be at its lowest possible value, which is in the range of 5 to 200 ohms.

16.14.3 If R_D is much greater than this, the N-channel resistance (r_{DS}) is sometimes regarded as negligible.

16.14.4 <u>Remember:</u>

[1] When the gate to source voltage of the JFET is 0 V ($V_{GS} = 0$), a closed switch is represented.

[2] In such a situation, the drain to source voltage (V_{DS}) is 0 V, or of a very low value.

16.14.5 <u>Remember:</u>

[1] As the gate source voltage of the JFET becomes more negative, and the resistance of the N-channel increases until the cut-off point is reached (at this point, the resistance of the channel is assumed to be infinite), an "open switch" results.

[2] In such a situation, $V_{DS} = V_{DD}$.

16.15 When a JFET is operated between the two extremes of current saturation and current cut-off, it offers variable resistance.

Review

17.1 ON and OFF are the terms used in digital electronics to describe the two transistor conditions mentioned above.

17.2 The similarity of a transistor to a mechanical switch is utilized in many digital situations.

17.3 When a transistor is being biased at a point falling between the ON and OFF conditions, it can be viewed as a variable resistance and utilized as an amplifier.

17.4 The voltage at the gate controls the drain current flow; this is similar to the base voltage that controls the collector current in a BJT.

17.5 The JFET is a high impedance device that does not draw current from the gate circuit.

17.6 The BJT is a device that requires some base current to operate it and has a relatively low impedance.

8 TRANSISTOR SWITCHES

1.1   Almost all industrial controls are presently transistor switches.

1.2   A computer consists entirely of transistor switches.

1.3   Computers make use of Boolean algebra.

1.4   Boolean algebra relies on the two logic states - TRUE and FALSE.

1.5   The logic states, TRUE and FALSE, can easily be represented electronically by a switch which is ON or OFF.

1.6   For the fast computations and manipulations of Boolean algebra, the transistor switch is an ideal device.

1.7   Transistors are found everywhere and play an important role in our daily lives.

2.1   To perform effectively as a CLOSED switch, the transistor switch's collector voltage has to be at the same voltage as its emitter.

2.2   In this instance the collector voltage will be at ground potential.

2.3   The transistor is thus said to be turned ON.

2.4   When the transistor is turned ON the collector voltage is 0 volt.

2.5   An ON transistor resembles a closed mechanical switch.

2.6   There will be a very small voltage drop across the transistor from the collector to the emitter, in actual practice.

2.6.1   This voltage drop is in fact a saturation voltage.

2.6.2   It is the smallest voltage drop that the transistor can have when it is ON as "hard" as possible.

2.6.3   We can consider this collector voltage to be a negligible value and to be 0 volt.

2.6.4   This is a good assumption for a quality switching transistor.

3.1   In calculating how to turn a transistor ON, the following steps can be used:-

[1] Determine the collector current.
[2] Check the value of β.
[3] From the results obtained by the above two steps, calculate the value of I_B.
[4] Calculate the value of R_B.
[5] Produce the final circuit.

3.2   The following is a figure of a transistor switch:-

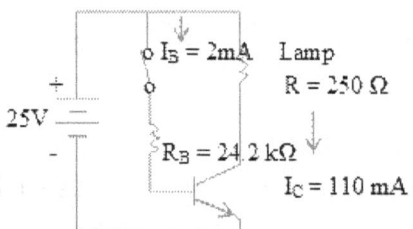

3.2.1 $I_B = I_C/\beta$, and, R_B = Lamp Voltage/I_B

4.1   In practice, if the supply voltage is much greater than the 0.7 V drop between the base and the emitter, the 0.7 V drop can be ignored.

4.2   In this instance, we can assume that all the supply voltage appears across the base resistor R_B.

4.3   This will simplify the arithmetic. (In any case, resistors are only accurate to within 10% of their stated value.)

4.4   The 0.7 V should not be ignored if the supply voltage is less than 10 volts.

Turning The Transistor Off

5.1   A transistor that is OFF resembles an open mechanical switch.

5.2   No current flows through the load, that is, no collector current flows, if the transistor is OFF, or OPEN.

5.3   When a transistor switch is OPEN, the two terminals are at different voltages (usually the supply voltage and the ground voltage), <u>unless</u> another resistor is across the switch.

5.4   <u>Remember</u>:
[1] When a transistor switch is OPEN no current flows.
[2] For a transistor to turn OFF and act like an OPEN switch, no base current is required.

5.5   No current flows when the transistor is OFF as "hard" as possible.

5.5.1   However, there is a very small leakage current flowing through the transistor.

5.5.2   As this leakage current is so low, it is often ignored in practice, especially in quality switching transistors.

5.5.3   For the different types of transistor, this leakage current is specified by the respective manufacturers (as is the case for the small voltage drop that occurs when the transistor is in the ON state).

5.6   To turn the transistor ON and make it behave like a closed mechanical switch, it is necessary to make the collector current flow.

5.7   This is done by providing the base current (and calculating the value of R_B that is required).

5.8 The base is tied to 0 V through a resistor R_2 in order to turn the transistor OFF, thereby ensuring that no collector current flows (making the transistor behave like an OPEN switch).

Reasons Why Transistors Are Used As Switches

6.1 A transistor can be used as a switch to turn a lamp current on and off.

6.2 Transistors are often interposed between a lamp and a mechanical switch.

6.3 But, this is not the main use for the transistor.

6.4 It is usually some other item or component in the base circuit that does the switching.

6.5 It is usually some other item in the collector circuit that is being controlled.

6.6 The following examples demonstrate the advantages that are gained by introducing the transistor switch into a circuit:-

6.6.1 Example One

A lamp is placed in a dangerous area, e.g., a radioactive chamber. In this instance, the switch has to be placed in a safe place. The transistor switch is the switch that is required as it provides extra isolation between the operator and the chamber.

6.6.2 Example Two

A high intensity lamp which requires much current is used. This current has to flow through the wires between the lamp and the switch. By using the transistor as the lamp switch and controlling the base current with a smaller switch in the base circuit, only the smaller base current will flow through the wires. For both the operator and the wiring this presents a much safer arrangement.

6.6.3 Example Three

Switching high current in wires produces interference in adjacent wires, and this can adversely affect communications or computer interconnections. By using the transistor switch, the current can be reduced to a smaller value, which is most desirable.

6.6.4 Example Four

Besides the lamp, some electronic control or communications equipment, could be switched on and off. In such an instant, the signal should not find its way into unnecessary places outside the equipment. It is much easier to utilize an outside switch to control the signal without the signal ever leaving the equipment. The transistor switch is ideal for this purpose.

Remember:

[1] A transistor's switching action can be directly controlled by an electrical signal, as well as a mechanical switch in the base circuit.

[2] This gives a great deal of flexibility to the design and allows for simple electrical control.

[3] Other advantages of the transistor switch include reduction of interference, remote switching control, safety and lower design costs.

6.6.5   Example Five

The ON and OFF times of mechanical switches are not very accurate, but that of a transistor switch can be very accurately controlled. In fact, transistors are much more accurate and controllable than any other device that is known. The transistor switch is ideal for applications such as photography where it is necessary to illuminate an object or expose a film for a precise period of time.

6.6.6   Example Six

A transistor switch can turn ON and OFF many millions of times a second, and can continue operating like this almost indefinitely. The transistor is virtually indestructible under normal operating conditions. Perhaps, the transistor is the longest lasting and most reliable component ever known, unlike mechanical switches which usually fail after a few thousand operations.

6.6.7   Example Seven

Much industrial control and communications information take the form of digital codes. Most digital codes rely on the presence or absence of a voltage rather than the level of the voltage. Transistor switches are ideal for controlling voltages. By using transistor switches very complex and very fast codes are possible.

6.6.8   Example Eight

The transistor also plays an important role in the medical field. For instance, the heart pacemaker has a transistor which causes it to pulse regularly over a long period of time, thereby keeping the wearer alive. This could never be done by a mechanical device.

6.6.9   Example Nine

With modern manufacturing techniques transistors can be miniaturized to such a great extent that many of them, thousands, can be integrated into a single device. These devices are called integrated circuits (ICs) and are extremely small. Transistors are the basis for the operation of many new products and are used practically everywhere. For example, the miniaturization of transistor switching circuits has made possible the home computer.

Remember: Additional advantages of the transistor switch are as follows:

[1] High speed operation
[2] Can be accurately controlled
[3] Long life

[4] Reliable operation

[5] Low power consumption

[6] Very small

[7] Can be manufactured with large numbers in an extremely small space

7.1 One transistor can be used to turn another transistor ON and OFF.

7.2 The second transistor can be used to operate a lamp or some other load.

7.3 All the switching can be carried out with low current switches, if many high current loads are to be switched.

7.4 It is possible for one remote switch to turn many items ON and OFF simultaneously.

The Two Transistor Switching Circuit

8.1 All modern electronic equipment comprises of multiple switching transistors, whereby one transistor is utilized to switch others ON and OFF.

8.2 The following is a typical two transistor switching circuit (The lamp is the load and the mechanical switch is the actuating element):-

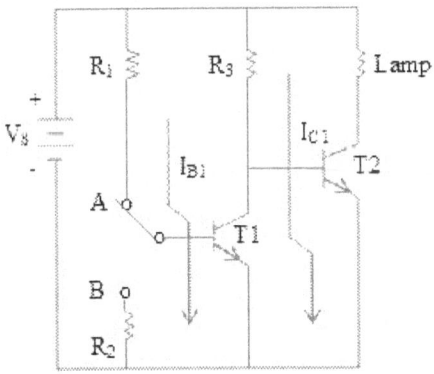

8.2.1 If the mechanical switch is in <u>position A</u>, base current I_{B1} will flow through R_1 into the base of T1.

8.2.2 This turns T1 ON, and causes the collector current I_{C1} to flow.

8.2.3 The collector voltage drops to 0 V.

8.2.4 As the base of T2 is joint to the collector of T1, the base of T2 drops to 0 V too.

8.2.5 This makes sure that T2 is turned OFF.

8.2.6 All of the collector current through R_3 flows as corrector current I_{C1} to ground through

T1.

8.2.7 As there is no base current I_{B2} for T2, it cannot turn ON.

8.2.8 The lamp therefore does not light.

8.3 Take a look at the same two transistor switching circuit again, but with the switch in position B now:-

8.3.1 When the mechanical switch is in position B, no base current flows into T1.

8.3.2 Therefore, no collector current flows, and T1 is OFF.

8.3.3 The current which flows through R_3 is all base current I_{B2} for T2.

8.3.4 This turns T2 ON.

8.3.5 The lamp is thus lighted.

The Three Transistor Switching Circuit

9.1 To extend the two transistor switching circuit a third transistor can be added.

9.2 The following figure is a three transistor switching circuit:-

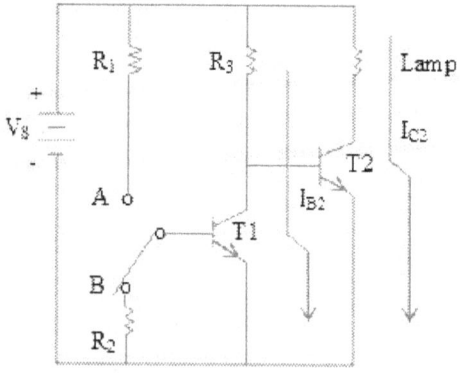

9.2.1 T1 will turn T2 ON and OFF.

9.2.2 T2 will operate T3.

9.2.3 (i) If the mechanical switch is in <u>position A</u>, T1 will turn ON.
(ii) T2 will be OFF.
(iii) The current through R_4 will flow into the base of T3.
(iv) T3 will turn ON.
(v) The lamp <u>will</u> light.

9.2.4 (i) If the mechanical switch is in <u>position B</u>, T1 will turn OFF.
(ii) T2 will be ON.

(iii) The current through R_4 will flow through T2 to ground.

(iv) T3 will turn OFF.

(v) The lamp <u>will not</u> light.

9.3   The switch position in the three transistor switch circuit is opposite that in the two transistor switch circuit.

9.4   The inclusion of an extra transistor to change a switch's operating state is quite common in electronics.

10.1   <u>Remember</u>:

[1] The transistor is much faster than the mechanical switch.

[2] The transistor can be more accurately controlled than the mechanical switch.

[3] The transistor is easier to operate remotely than the mechanical switch.

[4] The transistor is more reliable than the mechanical switch.

[5] Under normal conditions, transistors are indestructible. They do not wear out as they have no moving parts. As for the mechanical switch, it will fail after several thousand operations. Transistors will operate for years and will do so at several million times a second.

Switching The Junction Field Effect Transistor

11.1   The JFET is a "normally ON" device.

11.2   When no voltage is applied to the input terminal (the gate) the JFET is in a highly conductive state, offering low resistance to current flow.

11.3   With a voltage applied to the gate the JFET conducts less current as the drain to source channel resistance has increased.

11.4   As the voltage increases, the resistance can become so large that the JFET will cut out the flow of current at some point.

11.5   <u>Remember</u>:

[1] The three terminals of the JFET are the drain, source and gate.

[2] The gate controls the operation of the JFET.

[3] When the gate voltage is 0 V (at the same potential as the source), the drain current will be at its highest value (and the JFET will be ON).

[4] When the gate to source voltage is high, the drain current will be 0 A (and the JFET will be OFF).

Review

12.1   The transistor switch is a simple device.

12.2   It is necessary to calculate the resistor values when using transistors.

12.3   The transistor will burn out if the current is too large.

12.4   There is a maximum voltage the transistor will stand.

12.5   Transistors can switch ON and OFF at various speeds.

12.6   The JFET will not switch as fast as the BJT.

12.7   The JFET has an advantage over the BJT as it has a large input resistance.

12.8   The JFET does not draw any current from the control circuit in order to operate.

12.9   The control circuit of the JFET always has a large resistance.

12.10   The control circuit of the BJT has a small input resistance, when the BJT is in the ON state.

9 TRANSISTOR AMPLIFIERS

1.1   Transistors can be used to amplify small AC signals.

1.2   An amplifier, e.g., is required to amplify the signal from a microphone in order for it to be transmitted to the homes of listeners, and to amplify the output from a record pick-up that is too weak to drive a speaker.

1.3   Televisions, radios and hi-fis are some examples of modern electronic amplifying devices.

1.4   A one-transistor amplifier is capable of amplifying a small signal to a usable level.

1.5   Depending on the imagination of the circuit designer, many amplifier circuit configurations are possible.

1.6   There are several types of amplifier, which include the JFET, the BJT and the op-amp (operational amplifier), an integrated circuit.

2.1   The small voltage variations of the transistor will induce small variations in the base current, if a small AC signal is applied to its base.

2.2   These will be amplified by a factor of β.

2.3   There will thus be corresponding variations in the collector current.

2.4   These collector current variations will in turn bring about similar variations in the collector voltage.

2.5   Note: The β used in calculating AC gain is not the same as the β used in DC variation calculations.

2.6   When the transistor is operating properly, the AC β is the value of the common emitter AC forward current transfer ratio.

2.7   The AC β appears as h_{fe} in manufacturers' data sheets for individual transistors.

2.8   The AC β is utilized when we have to calculate the AC output for a given AC input or determine an AC current variation.

2.9   The DC β is used when there are calculations involving the collector and the base DC current values.

2.10   It is important for us to know which β to utilize - AC β or DC β.

2.11   Sometimes, the DC β is listed as h_{FE} or β_{dc}.

3.1   Look at the following circuit:-

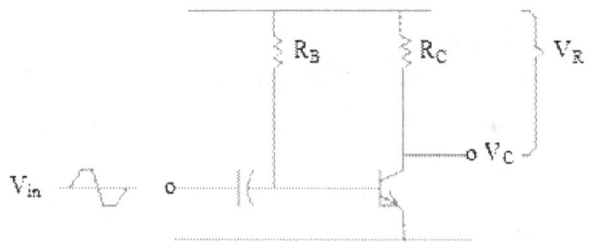

3.1.1   The base current in the above circuit increases as V_{in} increases.

3.1.2   This causes the collector current to increase.

3.1.3   The collector current increase brings about an increase in the voltage drop across R_C.

3.1.4 V_C thus decreases.

3.1.5   The capacitor at the input will easily pass AC (low reactance) and block DC (infinite reactance) signals.

3.1.6   This is a common filtering technique for AC inputs and outputs.

4.1   <u>Remember</u>:

         [1] If the input signal decreases, the collector voltage (V_C) will increase.

         [2] If a sine wave is applied to the input, an inverted sine wave would appear at the collector.

5.1   Refer to the following circuit:-

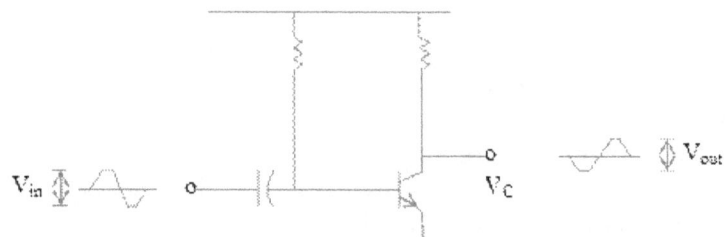

5.1.1   In the above circuit, V_{in}, the input voltage, is applied to the base. (The input voltage is actually applied between the base and the emitter.)

5.1.2   The voltage variations at the collector are centered around the DC bias point, V_C.

5.1.3   The voltage variations at the collector will be <u>larger</u> than the input voltage variations.

5.1.4   In other words, the voltage variations at the collector are <u>amplified</u>.

5.1.5   These amplified variations at the collector will subsequently be utilized to drive some components, e.g., a speaker.

5.1.6   These amplified variations at the collector are called the output voltage or output signal.

5.1.7   The AC output voltage (or output signal) is indicated by V_{out}, to differentiate these AC output variations from the DC bias level.

5.1.8   More often than not, the AC output voltage is expressed as a peak-to-peak value.

6.1   Remember:
        [1] V_C is the collector DC voltage, or the bias point.
        [2] V_{out} is the AC output voltage.

7.1   The voltage gain of the amplifier is the ratio of the output voltage to the input voltage.

7.2   The following formula is used to calculate the voltage gain:-

$$\text{Voltage Gain} = A_V = V_{out}/V_{in} = \beta \times R_L/R_{in}$$

7.3   Measuring the AC voltage in and out with an oscilloscope can directly give the voltage gain.

7.3.1   When trying to find the voltage gain, look at the AC peak-to-peak voltages only.

7.3.2   Ignore the DC levels.

7.3.3   Remember:
        [1] Gain is an AC concept.
        [2] Gain has absolutely nothing to do with the DC bias point (or collector DC voltage).

7.4   The voltage gain cannot be guaranteed from one transistor amplifier to another.

7.4.1   The voltage gain is very "temperature dependent", as both β and R_{in} change greatly with temperature. (For instance, a transistor amplifier designed to work in the very hot desert would fail completely in the very cold Arctic, and, a transistor amplifier designed to work indoors would fail to work outdoors on either a cold or hot day.)

7.4.2   A transistor amplifier whose voltage gain (and DC bias point) change in this manner is said to be "unstable".

7.5   A transistor amplifier should be as stable as possible in order to operate efficiently.

How To Overcome Amplifier Instability

8.1   Three changes are made to the basic transistor amplifier circuit in order to overcome the instability.

8.1.1   To ensure the DC bias point stability, two resistors, R_1 and R_2, are added in, as shown below:-

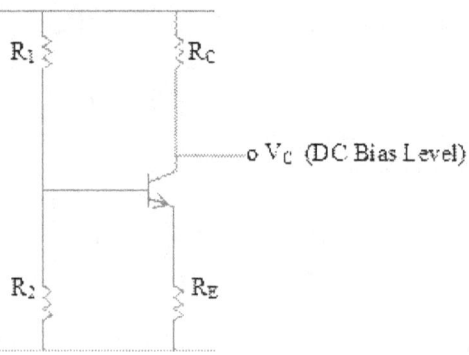

8.1.2 An emitter resistor, R_E, is added in to ensure the AC gain stability.

8.1.3 The same circuit is shown below with the important DC currents and voltage shown:-

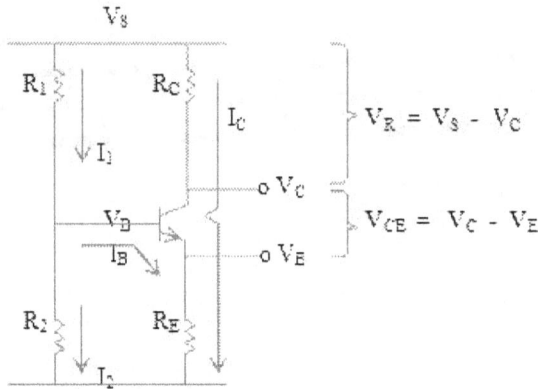

8.1.4 In designing an amplifier circuit and choosing the resistor values, two things have to be achieved, which are as follows:-

[1] A stable DC bias point.
[2] A stable AC gain.

8.1.5 The formula that can be used to calculate the voltage gain is as follows:-

$$A_V = V_{out}/V_{in} = R_C/R_E$$

(Note: The AC gain has now been made independent of the transistor β and the input impedance.)

8.1.6 Remember: The β and the input impedance of the transistor vary from transistor to transistor and with temperature.

8.1.7 By including the three resistors, R_1, R_2 and R_E, in the above circuit, the AC gain will be constant regardless of all the above variations.

8.1.8 For example, in the above circuit, if R_C = 10 kΩ and R_E = 1 kΩ, the AC voltage gain will remain constant at 10 regardless of the value of β (whether 100, 600, 900, and so on).

Biasing

9.1 Biasing the transistor amplifier circuit is setting its DC output voltage level.

9.2 Look at the circuit below:-

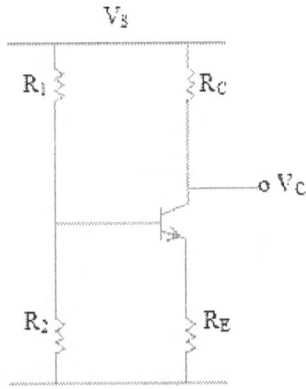

9.2.1 In the above circuit, resistors, R_1, R_2 and R_E, will bias the circuit to its correct DC conditions and provide the gain that is required.

9.2.2 To design the biased transistor amplifier circuit, the following three factors have to be stated as desired conditions:-

[1] The AC gain that is required.
[2] The DC output voltage level.
[3] The value of the collector resistor.

9.2.3   The following is the design procedure:-

[1] Find R_E. (Use $A_V = R_C/R_E$)
[2] Find V_E. (Use $A_V = V_R/V_E = (V_S - V_C)/V_E$)
[3] Find V_B. (Use $V_B = V_E + 0.7$ V)
[4] Find I_C. (Use $I_C = (V_S - V_C)/R_C$)
[5] Find I_B. (Use $I_B = I_C/\beta$)
[6] Find I_2. (Refer to the circuit. Let I_2 be 10 I_B.)
[7] Find R_2. (Use $R_2 = V_B/I_2$)
[8] Find R_1. (Use $R_1 = (V_S - V_B)/(I_2 + I_B)$)
[9] Choose the nearest standard values for resistors, R_2 and R_1, as steps [7] and [8] usually give non-standard values for the resistors.
[10] Use the voltage divider formula to see if the standard values that were chosen in step [9] above give a voltage level close to V_B which was found in step [3]. (The voltage level should be within 10% of the ideal.)

9.3   The above procedure will give a transistor amplifier (which will have the required DC voltage levels and AC gain) that will work.

9.4   The following is the actual gain formula for the transistor amplifier:-

$A_V = \beta \times R_C/(R_{in} + R_E)$ (As the collector resistor is the total load on the amplifier, R_C is used instead of R_L.)

9.5   Two transistor amplifiers can be <u>cascaded</u> to obtain larger voltage gains.

9.6   In cascading, the output of the first transistor is fed into the input of the second transistor.

9.7   Below is a circuit which has two transistor amplifiers in cascade:-

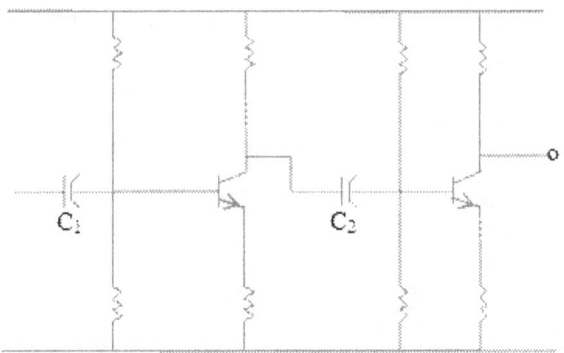

9.7.1   The two transistor amplifiers in the above circuit are designed individually.

9.7.2   They are connected by capacitor C_2.

9.7.3   Capacitor C_2 has a very low reactance at the lowest frequency required to be passed.

9.7.4   The overall gain is obtained by multiplying the individual gains.

9.7.5   In the above circuit, if the first amplifier has a gain of 12 and the second has a gain of 15, then the overall gain is 180.

9.7.6   Very large gains can be obtained if the capacitors are placed in parallel with each emitter resistor.

9.8   It is not uncommon to have gains of several thousand.

9.9   To obtain maximum gain and to adjust the output voltage with a variable resistor, a bypassed emitter can be utilized.

9.10   In the following figures two bypassed emitters are shown:-

9.10.1 Method [A] is the most common.

9.10.2 Method [A] is often found in consumer equipment as the "volume control".

The Emitter Follower

10.1   The following figure is an emitter follower, sometimes called the common collector amplifier:-

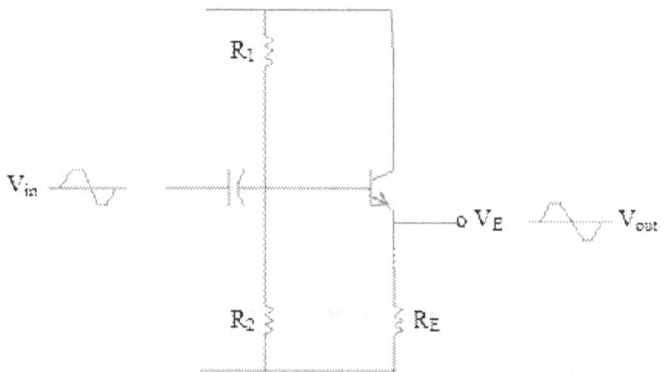

10.1.1   In the above circuit, the output signal is taken from the emitter.
10.1.2   There is no collector resistor.
10.1.3   In the above circuit, the peak-to-peak value is almost the same as the input signal.
10.1.4   The gain of the circuit is slightly less than 1. (It is often considered to be 1 in practice.)
10.1.5   The phase of the output is the same as the phase of the input.
10.1.6   The output is <u>not</u> inverted.
10.1.7   The output is considered to be the same as the input.
10.1.8   The output has a very high input resistance.
10.1.9   This output signal draws very little current from the signal source.
10.1.10 This output also has a very low output resistance.
10.1.11 The signal at the emitter seems to be coming out from a battery or signal generator which has a very low internal resistance.
10.2   <u>Remember</u>:
        [1] The voltage gain of an emitter follower is 1.
        [2] The output of the emitter follower is <u>not</u> inverted.
        [3] The input resistance of the emitter follower is <u>high</u>.
        [4] The output resistance of the emitter follower is <u>low</u>.
10.3   The following example shows the importance of the emitter follower circuit. Take a look at the circuit below:-

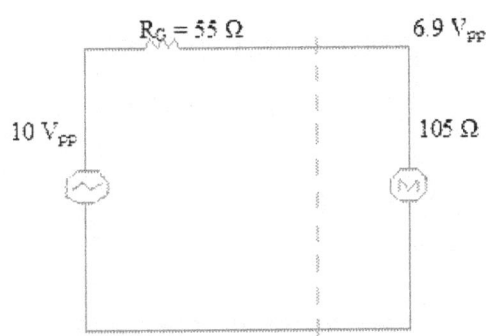

10.3.1 A small AC motor that has 105 ohms resistance is driven by a 10 V_{pp} signal from a generator.

10.3.2 The signal generator has an internal resistance.

10.3.3 The voltage across the motor is about 6.9 V_{pp}.

10.3.4 Next, place an emitter follower between the motor and the generator, as shown below:-

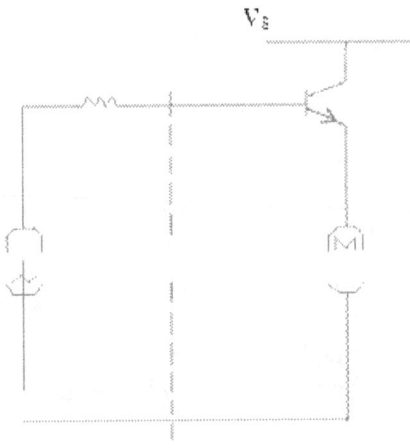

10.3.5 The following formula is used to calculate the emitter follower's input resistance:-

$$R_{in} = \beta \times R_E$$

10.3.6 The emitter follower now provides the load on the generator instead of the motor.

10.3.7 The load on the generator is now 10,500 ohms instead of 105 ohms.

10.3.8 This increase in ohmage will not cause the voltage of the generator to fall.

10.3.9 The generator voltage will remain at 10 V_{pp}.

10.3.10 The voltage of the emitter will also remain at 10 V_{pp}.

10.3.11 The current through the motor is now produced by the power supply and not the generator.

10.3.12 The transistor behaves like a generator which has a very low internal resistance.

10.3.13 This internal resistance is actually the emitter follower's output impedance.

10.3.14 The following formula is used to calculate this internal resistance:-

$$R_o = \text{Internal Resistance Of Generator}/\beta$$

10.3.15 In the above circuit, this internal resistance is about 0.5 ohm.

10.3.16 In the above circuit, there is a load of 105 ohms connected to a generator which has an internal resistance of 0.5 ohm only.

10.3.17 In this way, the output voltage of 10 V_{pp} is maintained across the motor.

10.4 <u>Remember</u>:

[1] The emitter follower in the above circuit is used to drive a load which could not be driven directly or directly connected to a generator.

[2] The emitter follower is important because of the two following properties:-

(i) It has a high input resistance.

(ii) It has a low output resistance.

10.5 The design procedure for the emitter follower is the same as that for the biased transistor amplifier, which was brought up earlier.

10.5.1 <u>Note the following points when designing the above circuit</u>:

[1] Usually half the supply voltage is chosen, and V_E is a DC voltage level.

[2] R_E is often a given factor, especially if it is a meter or a motor which has to be driven.

[3] When calculating R_1 with the equation, $R_1 = (V_S - V_B)/(I_2 + I_B)$, I_B in the equation is usually ignored.

10.6 The emitter follower in the above circuit is biased by the use of resistors (R_1, R_2 and R_E).

10.6.1 The emitter follower does not depend on the value of R_E, as V_E is set by the biasing resistors.

10.6.2 In this circuit, any value of R_E can be utilized.

10.6.3   The following is used to calculate the minimum value for R_E:-

$$R_E \; = \; 10 \; R_2/\beta$$

(Note: R_E has a wide range of values.)

10.6.4   Exact bias resistor values are not required.

10.6.5   All these make the circuit design task an easier one.

Analyzing The Amplifier

11.1   A circuit which has been designed can be "analyzed".

11.2   "Analyzing" here means calculating the collector DC voltage (the bias point) and finding the AC gain.

11.3   When analyzing the amplifier circuit, the following procedure, which is basically the reverse of the design procedure, can be used:-

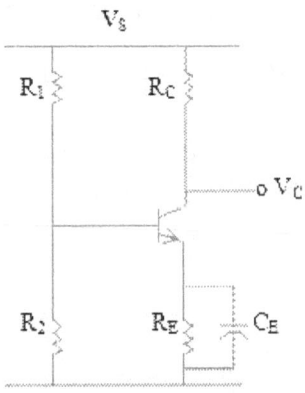

[1]   Find V_B. (Use $V_B \; = \; V_S \; x \; (R_2/(R_1 \; + \; R_2))$

[2]   Find V_E. (Use $V_E \; = \; V_B \; - \; 0.7 \; V$)

[3]   Find I_C. (Use $I_C \; = \; V_E/R_E$. Note: $I_C \; = \; I_E$)

[4]   Find V_R. (Use $V_R \; = \; R_C \; x \; I_C$)

[5]   Find V_C, the <u>bias point</u>. (Use $V_C \; = \; V_S \; - \; V_R$)

[6]   Find A_V, the <u>AC voltage gain</u>. (Use $A_V \; = \; R_C/R_E$ or $A_V \; = \; \beta \; x \; R_C/R_{in}$. When using the second formula, the value of R_{in} (or h_{ie}) should be obtained from the manufacturer's data sheets for the transistor.)

The JFET Amplifier

12.1　The figure below is a typical biased JFET:-

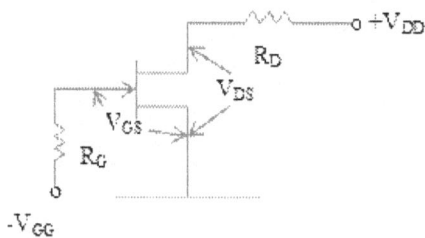

12.1.1　The above is an N-channel JFET.

12.1.2　V_{GS} is the gate to source voltage.

12.1.3　V_{GS} should be 0 V to turn the JFET "hard ON".

12.1.4　When V_{GS} is 0 V and the JFET is "hard ON", the drain saturation current (I_{DSS}) will flow.

12.1.5　To turn the above N-channel JFET "hard OFF", V_{GS} should be a large negative voltage. (V_{GS} should be larger than or equal to the cutoff voltage.)

12.1.6　In changing V_{GS} between the two extremes, 0 V and large negative voltage, the JFET is operated as a switch.

12.2　If the JFET is operated with a gate to source voltage that is about halfway between the ON and OFF states, the drain current that would flow can be calculated with the following equation:-

$$I_D = I_{DSS} (1 - (V_{GS}/V_P))^2$$

12.3　<u>Note</u>: [1]　For a given JFET, manufacturers give I_{DSS} and $V_{GS\ (off)}$ a wide range.
　　　　　[2]　Sometimes, typical values are given and they can be used as a starting point.
　　　　　[3]　Sometimes, I_{DSS} and $V_{GS\ (off)}$ are measured instead.

12.4　To turn the JFET into an amplifier, an input signal to the gate is necessary.

12.4.1　Applying an input signal to the gate brings about voltage changes on the gate that result in corresponding changes in drain current.

12.4.2　A sine wave of 0.5 V_{pp}, e.g., can be applied to the gate as shown below:-

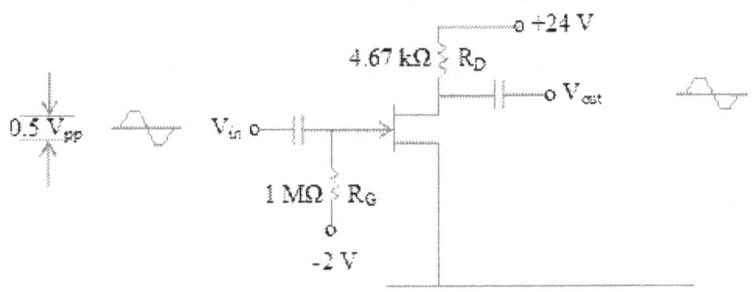

12.4.3 The gate voltage in the above circuit will vary from -1.75 V to -2.25 V.

12.4.4 The corresponding drain currents will be as follows:-

   [1] For V_{GS} = -1.75 V, I_D = 3.8 mA
   [2] For V_{GS} = -2.25 V, I_D = 2.3 mA

12.4.5 These drain currents will cause a change in the voltage drop across resistor R_D.

12.4.6 The variation in the voltage drop across resistor R_D will be as follows:-

   [1] For I_D = 3.8 mA, V_{RD} = 17.7 V
   [2] For I_D = 2.3 mA, V_{RD} = 10.7 V
      Voltage Variation = 17.7 V - 10.7 V = 7 V_{pp}

12.4.7 Also, the output voltage will vary.

12.4.8 The output voltage variation will be as follows:-

   [1] For I_D = 3.8 mA, V_{out} = 24 V - 17.7 V = 6.3 V
   [2] For I_D = 2.3 mA, V_{out} = 24 V - 10.7 V = 13.3 V
      Output Voltage Variation = 13.3 V - 6.3 V = 7 V_{pp}

12.5 The voltage gain for this JFET amplifier can be obtained with the following formula:-

$$A_V = -V_{out}/V_{in}$$

where: the negative sign in -V_{out} shows 180^0 phase inversion from the input to the output.

12.6   Note: [1] As the input voltage on V_{GS} decreases (becomes more negative), the output voltage <u>increases</u>.

        [2] As the input voltage on VGS increases (toward 0 V), the output voltage <u>decreases</u>.

12.7   The following expression can also be used to calculate the voltage gain of this JFET amplifier:-

$$A_V = -(g_m)(R_D)$$

where: g_m is called the transconductance (and also the forward transfer admittance), a dynamic characteristic of the JFET.

(<u>Note</u>: Manufacturers' specification sheets generally provide a typical value for g_m for each JFET.)

12.8   The following formula can be used to calculate g_m, the transconductance:-

$$g_m = \Delta I_D / \Delta V_{GS}$$

where: [1] ΔV_{GS} indicates the variation or change in V_{GS}.

        [2] ΔI_D indicates the variation or change in the corresponding drain currents.

(<u>Note</u>: Siemens or mhos is the unit for transconductance, which is a very small number.)

12.9   A one power supply and a self-biasing technique can also be used to operate the JFET.

12.9.1 Such a circuit is shown below:-

(<u>Note</u>: The voltage drop across R_S becomes the gate to source voltage as there is no current flow at the gate.)

The Operational Amplifier (Op-amp)

13.1   The op-amp which is used today is really an integrated circuit (IC).

13.2   The device has numerous solid state components and resistors which are constructed on the same silicon substrate and are in a form which is very small and easily utilized.

13.3   Op-amps are available in different case configurations, e.g., the flat pack, the mini DIP, the 14 pin DIP and the TO metal package.

13.4   Two op-amps (dual) or four op-amps (quad) can be found in the same IC package.

13.5   A typical op-amp may have 20 or more transistors in its design.

13.6   Op-amps are often considered a circuit device or component in themselves.

13.7   Today, op-amps are very common.

13.8   Op-amps very closely resemble an ideal amplifier in characteristics.

13.9   The characteristics of, or, advantages of using, op-amps are as follows:-

[1] They are cheap.

[2] They have a wide range of applications.

[3] They are small in size.

[4] They have high gain (and can thus be used for amplifying small signal levels).

[5] They have low output impedance (and are not affected by the load).

[6] They have high input impedance (and do not require input current).

13.10   Large numbers of transistors and components are found in an op-amp.

13.10.1 All these can be packed into a small IC package and a lot of space is saved in this way.

13.11   The following is the general symbol for the op-amp:-

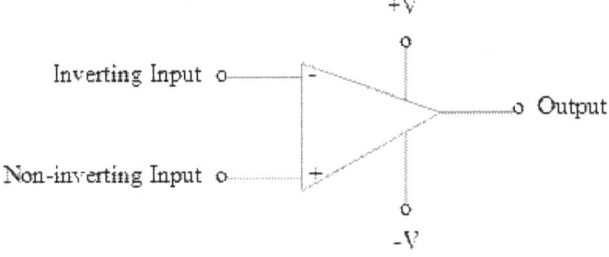

13.11.1 An input at the non-inverting input will bring about an output that is <u>in phase</u> with the input.

13.11.2 An input at the inverting input will bring about an output that is <u>$180°$ out of phase</u> with the input.

13.11.3 Both positive and negative supplies are required.

13.11.4 The values of both the positive and negative supplies will be specified for the particular op-amp that is used.

13.11.5 Sometimes, other terminals are shown for different applications.

13.11.6 In order to make the op-amp and associated circuitry perform the function that is required, the external components and connections have to be arranged.

13.12   The electronic circuit designer can refer to the numerous available handbooks and applications manuals which describe op-amp circuits.

13.13   <u>Remember:</u>

[1] The op-amp has five terminals: (a) Two input terminals. (b) One output terminal. (c) Two power supply terminals.

[2] When the input is connected to the inverting input, the output is <u>$180°$ out of phase</u> with the input.

[3] When the input is connected to the non-inverting input, the output is <u>in phase</u> with the input.

13.14   The circuit below, which is the most basic, utilizes a type 741 op-amp as an inverting amplifier:-

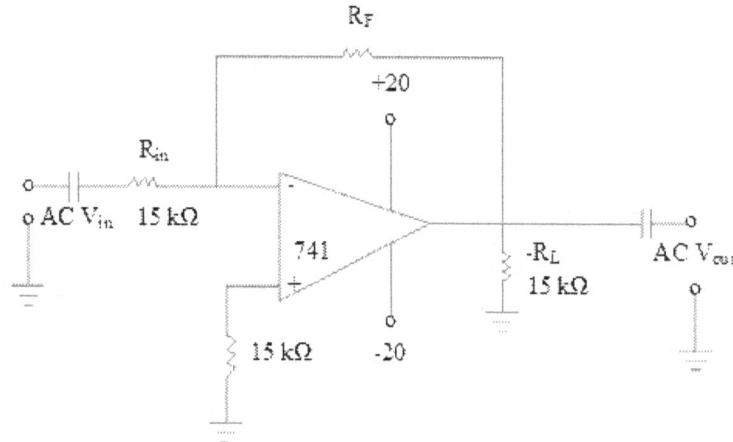

13.14.1 The following equation is used to calculate the voltage gain for the above circuit;-

$$A_V = -R_F/R_{in}$$

13.14.2 The feedback resistor, R_F, forms a feedback path from the output to the input. (Feedback loops are used by many op-amps.)

13.14.3 It is easy to saturate the op-amp (at maximum gain) with only very slight changes in potential between the two input terminals, as the op-amp has a high gain.

13.14.4 The op-amp is able to operate at lower gains over a wider range of input differences because of the feedback loop.

13.14.5 The feedback resistor value is chosen for achieving a certain voltage gain.

13.14.6 The capacitors in the above circuit keep DC voltages out of the input and output.

Review

14.1   The most common types of amplifiers that are in use today are the common emitter BJT, the common source JFET and the op-amp.

14.2   Amplification can be carried out by many variations and kinds of amplifiers, and other types of devices.

10 OSCILLATORS

1.1   An oscillator is a circuit that produces a continuous output signal.

1.2   An oscillator can be called a signal generator.

1.3   An oscillator circuit that produces a sine wave signal of constant amplitude and frequency is called a sine wave generator.

1.4   Sine waves include the 120 V AC (or 240 V AC in some countries) from the wall plug, TV and radio signals that are transmitted through the air, and many test signals that are used in electronics.

2.1   In practice, many types of oscillator circuits are used extensively.

2.2   Basic sine wave oscillators rely on the LC tuned circuit.

2.3   Various output signals are produced by other types of oscillators, e.g., signals like triangle waves, square waves and pulses.

3.1   The oscillator comprises of three aspects, viz., the amplifier, the feedback connections and the frequency determining components.

3.2   Using a resonant LC circuit as the frequency determining components and incorporating feedback from the output to the input can turn an amplifier into an oscillator.

3.3   An amplifier can replace a switch in a circuit.

3.4   Amplification is <u>necessary</u> in an oscillator.

3.5   Below is a typical oscillator circuit:-

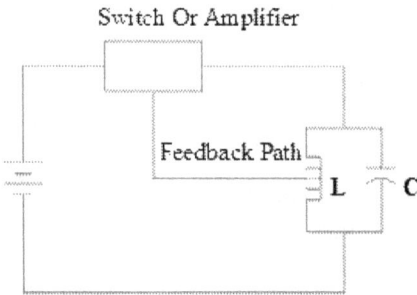

3.6   Feedback is achieved by connecting an amplifier's output to its input.

3.7   The circuit is said to have <u>negative</u> feedback (NFB) if the output fed back is <u>out of phase</u> with the input, as shown below:-

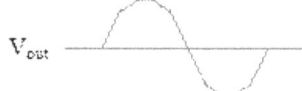

3.8   Connecting the collector of a transistor amplifier to its base through a feedback resistor (R_f) would produce negative feedback, as shown in the circuit below:-

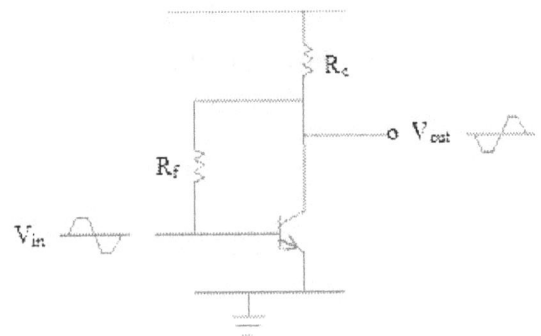

3.9   Negative feedback will <u>stabilize</u> the operation of an amplifier.

3.10   Negative feedback does the following:-

[1] It prevents the DC bias point drifting with component and temperature changes.

[2] It prevents changes in gain that are caused by component and temperature changes, and also reduces the gain (and makes it easier to control the gain).

[3] It reduces distortion in amplifiers, therefore improving the sound "quality".

3.11   High quality hi-fi amplifiers <u>always</u> have <u>negative feedback</u>.

4.1   If the feedback from the output is <u>in phase</u> with the input, the circuit is said to have <u>positive</u> feedback, as shown below:-

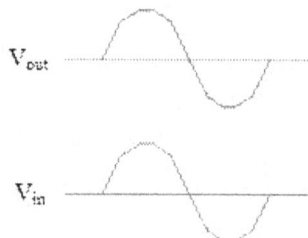

4.2 Connecting the collector of the second transistor in a two-transistor amplifier to the base of the first transistor causes <u>oscillations</u>.

4.3 Study the two-transistor circuit below:-

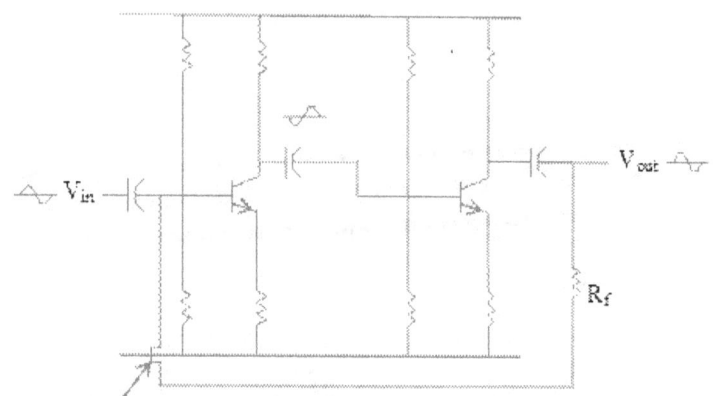

There is no connection made here

4.3.1 The above two-transistor circuit has <u>positive feedback</u>.

4.3.2 Positive feedback causes an amplifier to <u>oscillate</u> even when there is <u>no external input</u>.

4.4 <u>Remember</u>:
   [1] <u>Negative feedback</u> is used to <u>stabilize</u> an amplifier.
   [2] <u>Positive feedback</u> (PFB) is used in <u>oscillators</u>.
   [3] Connecting the output of an amplifier to its input produces feedback.

4.5 Look at the common emitter amplifier circuit, which has a simple resistive load (R_c), below:-

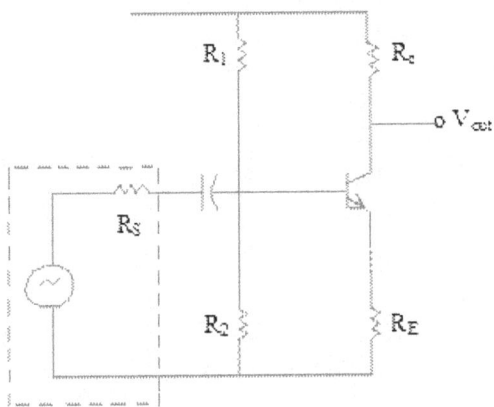

4.5.1   Negative feedback on the above amplifier would:-

[1] Stabilize it.
[2] Reduce the gain.
[3] Reduce distortion.

4.5.2   Positive feedback would cause the above amplifier to oscillate.
4.5.3   An amplifier is required in an oscillator circuit because the oscillator circuit has to have a large gain.
4.5.4   In the above circuit, if an external signal is applied to the base, the circuit will be amplified.
4.6   The basic formula for the gain of an amplifier is given below:-

$A_v = \beta \times R_L/R_{in}$

4.7   However, the gain formula for the above amplifier is as follows:-

$A_v = R_L/R_E = R_c/R_E$

(R_c is the only load in the above circuit.)

5.1   The figure below is the common base amplifier circuit, where there is an input to the emitter instead of the base:-

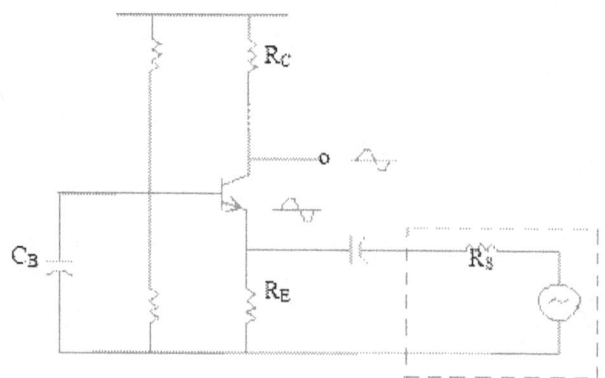

5.1.1   For the above circuit, the gain formula is still similar to the basic amplifier gain formula.

5.1.2   However, for the above circuit, when the signal is fed into the emitter, the input impedance to the amplifier is very low.

5.1.3   The basic gain formula is as follows:-

$$A_V = R_L/R_S$$

where: R_S is the output resistance or the source or generator impedance.

5.1.4 R_S above may also be regarded as the internal impedance of the source.

5.1.5   However, the actual gain formula for the above circuit is as follows:-

$$A_V = R_C/R_S \qquad \text{(as } R_C \text{ is the load given in the circuit)}$$

5.1.6   In the above circuit, the input and the output sine waves are all in phase.

5.1.7   The signal is not inverted although it is amplified.

5.1.8   Remember:

[1] The input signal to the amplifier, when applied to the emitter instead of the base, is amplified and not inverted.

[2] Compared to the common emitter amplifier, the input impedance of the common base amplifier is low.

[3] The gain formula for the common base amplifier is as follows:-

$$A_V = R_L/R_S$$

6.1   The above amplifier circuit can be modified by including a parallel LC circuit, as shown below:-

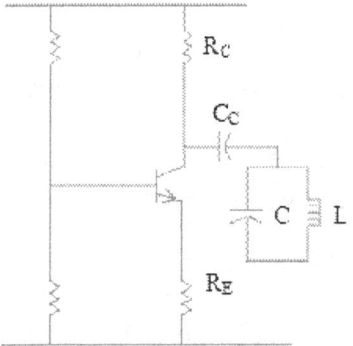

6.1.1   The above parallel LC circuit is sometimes called a tuned (or resonant) load.
6.1.2   To avoid a DC connection between the collector and the inductor, component C_C is added.
6.1.3   The inductor has a very small DC resistance.
6.1.4   The very small DC resistance of the inductance reduces the collector DC voltage to almost 0 V and prevents the circuit from behaving as an amplifier.
6.1.5   Remember:
        [1] The load in the above circuit is tuned or resonant.
        [2] An oscillator should have the following <u>three</u> aspects:-

        (i)   The amplifier.
        (ii) The frequency determining components.
        (iii) The feedback connections.

6.1.6   <u>Note</u>: One of these three aspects, the feedback connections, is missing from the above circuit.
6.1.7   The total load of the above circuit consists of the parallel combination of the collector resistor R_C and the tuned LC circuit.
6.1.8   The value of the total load will depend on the input frequency.
6.1.9   The LC circuit has a very high impedance, at the resonant frequency of the tuned load.
6.1.10 The value of R_C should be less than the impedance of the tuned circuit at resonance.
6.1.11 At the resonant frequency, the tuned load becomes about the same as R_C.
6.2   Remember:

[1] The gain formula for the common emitter circuit is: $A_V = R_C/R_E$

[2] The gain formula for the common base circuit is: $A_V = R_C/R_S$

6.3   Both the common base and the common emitter circuits are utilized in oscillators.

6.3.1   An extra capacitor is usually included in each case.

6.3.2   In the case of the common emitter amplifier, an emitter capacitor (C_E) is added.

6.3.3   In the case of the common base amplifier, a capacitor (C_B) is utilized to connect the base to ground.

6.3.4   In both cases, C_E and C_B will each bring about an increase in the gain of the amplifier.

6.3.5   The amplifier gain is increased to the point where it is <u>large enough</u> to allow the amplifier to be used as an oscillator.

7.1   The resonant frequency of an oscillator's LC circuit is the frequency at which the oscillator will oscillate.

7.2   The following formula is utilized to calculate the resonant (or oscillation) frequency:-

$$f_r = 1/(2\pi\sqrt{LC})$$

7.3   <u>Note</u>: [1] The calculated frequency, in practice, is never exactly the same as the actual measured frequency.

               [2] The same applies to the inductor and capacitor values.

               [3] The frequency will be affected by other stray capacitances in the circuit.

7.3.1   For the sake of accuracy in operation, an adjustable inductor or capacitor is utilized.

7.4   The formula for the impedance of the circuit at the resonant frequency is as follows:-

$$Z = 1/CX_r$$

where: $_r$ is the coil DC resistance.

7.5   The gain of the circuit at the resonant frequency is <u>large</u>, and is calculated with the following formula:-

$$A_V = Z/R_E$$

Feedback

8.1   To turn an amplifier into an oscillator, it is necessary to <u>feed back</u> a portion of the output signal to the input.

8.2   In order to cause oscillations, the feedback signal has to have the <u>correct phase</u>.

8.3   Usually, the feedback is taken from the tuned load.

8.4   The following are the three main types of oscillators used (each is named after the original inventor):-

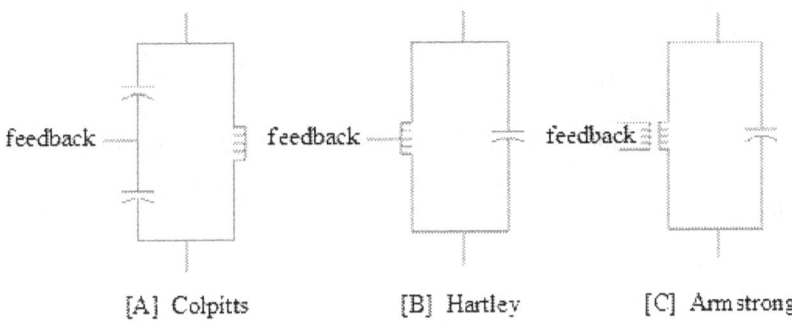

[A] Colpitts   [B] Hartley   [C] Armstrong

8.4.1   Each of the three oscillators above is a variation on the voltage divider circuit.
8.4.2   The Colpitts oscillator utilizes a capacitive voltage divider. (Remember: Colpitts and Capacitor each begins with a C.)
8.4.3   The Hartley oscillator utilizes a "tap" on the coil to give an inductive voltage divider.
8.4.4   The Armstrong oscillator utilizes a transformer (an extra winding, which usually has fewer turns than the main winding) rather than a true voltage divider.
8.5   In all the above three oscillators, between one tenth and one half of the output have to be used as feedback.
8.6   Remember:
       [1] In the Colpitts oscillator, the feedback is taken from a capacitive voltage divider.
       [2] The Hartley oscillator uses a tap on the coil for the feedback voltage.
       [3] The Armstrong oscillator uses a transformer instead of a voltage divider.
8.7   Look at the circuit below:-

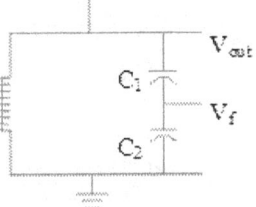

8.7.1   At one end of the above circuit there is the output voltage.
8.7.2   The other end of the circuit is effectively at ground.
8.7.3   The feedback voltage (V_f) is obtained between the junction of the two capacitors and ground.

8.7.4   The voltage divider formula used to calculate the feedback voltage, V_f, is as follows:-

$$V_f = V_{out}X_{C2}/(X_{C1} + X_{C2})$$

which becomes $V_f = V_{out}C_1/(C_1 + C_2)$

8.7.5   To find the resonant frequency of the above circuit, it is first necessary to find the equivalent total capacitance, C_T, of the two series capacitors in the circuit.

8.7.6   The formula for C_T is as follows:-

$$C_T = C_1C_2/(C_1 + C_2)$$

8.7.7   The resonant frequency formula for the Colpitts oscillator is as follows:-

If Q is less than 10, $f_r = 1/(2\pi\sqrt{LC_T})$

(The other formulae are as follows:-

$$f_r = 1/(2\pi\sqrt{LC}) \times \sqrt{1 - (r^2C/L)} = 1/(2\pi\sqrt{LC}) \times \sqrt{Q^2/(1 + Q^2)})$$

9.1   In the circuit below, the feedback voltage is obtained from a tap on the coil:-

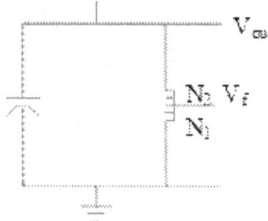

9.1.1   In the above circuit, N_1 and N_2 represent the number of turns in each of two sections of the coil.

9.1.2   The voltage divider formula, which is modified to include the number of turns in each part of the coil, used to find the voltage out of the above circuit is as follows:-

$$V_f = V_{out} \times (N_1/(N_1 + N_2))$$

9.1.3   The practical problem here is that often neither the number of turns nor the "turns ratio" is known as the exact place of the tap may not be known. (However, the number of turns in the coil is usually specified by the manufacturer.)

10.1   For the Armstrong oscillator, the feedback is obtained from the secondary winding in a transformer.

11.1   For all of the above cases, the voltage fed back from the output by the voltage divider is a small percentage of the total output voltage. (This V_f fraction is always below 1.)

11.2   The product of the amplifier gain and the feedback has to be greater than 1 in order to ensure oscillations. (A_V x V_f > 1)

11.2.1   This is usually easy to achieve as A_V is much greater than 1.

11.3   The external input is not applied to the oscillator.

11.4   The small part of the output signal is fed back to the oscillator's input.

11.5   The oscillations will commence spontaneously and will continue as long as power is supplied to the circuit, if this feedback is of the <u>correct amplitude and phase</u>.

11.6   This feedback is amplified by the transistor amplifier in order to sustain the oscillations.

11.7   The transistor amplifier converts the DC power from the battery or power supply into the AC power of the oscillations.

11.8   <u>Remember</u>:

[1] A tuned circuit with feedback of the correct phase and amplitude will turn an amplifier into an oscillator.

[2] An amplifier does not require an input to become an oscillator. (Oscillations will occur spontaneously if the feedback is correct.)

The Colpitts Oscillator

12.1   Amongst the LC oscillators, the Colpitts oscillator is the easiest to build.

12.2   Below is a typical Colpitts oscillator:-

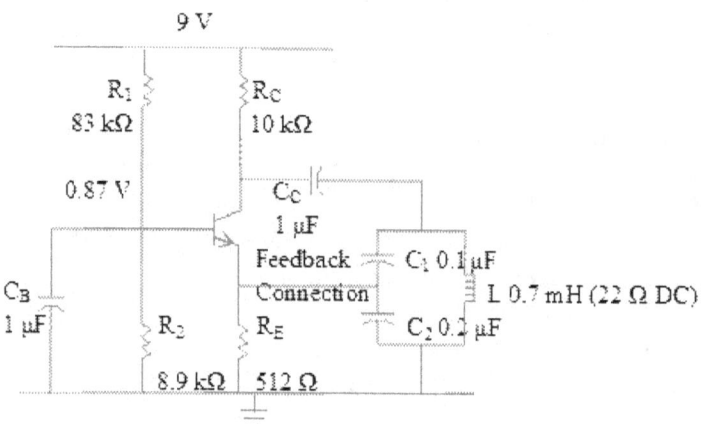

12.2.1   The capacitive voltage divider in the above circuit provides the feedback to the emitter.

12.2.2   The feedback has to be of the correct phase.

12.2.3   At the lowest frequency that is to be sustained in the oscillator, the reactance of capacitor C_B has to be low enough so that the AC signal sees the capacitor as a path, rather than R_2.

12.2.4   A frequency of 1 kHz and a reactance of 160 ohms can be chosen for a low-range oscillator.

12.2.5   The reactance at the frequency of oscillation should be checked.

12.2.6   If the value of R_2 is below 1.6 k Ω, then select a value of X_{CB} which is less than one-tenth the value of R_2.

12.2.7   The value of C_B should be large enough to present a low reactance bypass which will be enough to keep the DC bias point constant.

12.2.8 C_B could be <u>estimated</u> as follows:-

$$X_C = 160 \text{ ohms} = 1/(2 \times \pi \times 10^3 \times C_B)$$

Therefore, $C_B = 1 \ \mu F$.

C_B can be 1 μF or larger.

The Hartley Oscillator

13.1   In the Hartley oscillator, the feedback is obtained from a tap on the coil.

13.2   The wire lead colors in the Hartley oscillator are shown as G (green), B (black) and W (white).

13.3   The Hartley oscillator is shown below:-

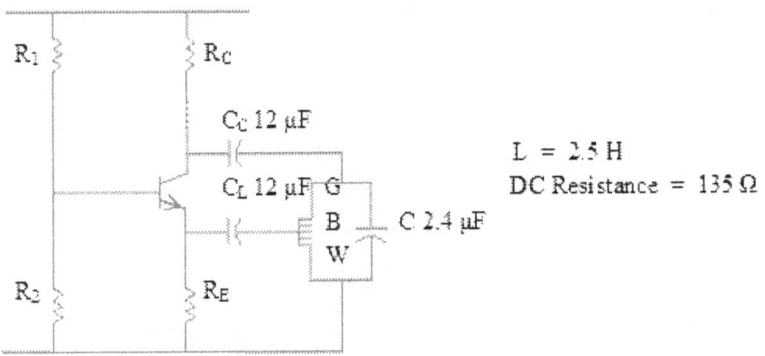

13.3.1 Capacitor C_L in the above circuit prevents the low DC resistance of the coil from pulling the emitter DC voltage down to 0 V.

13.3.2 At the oscillating frequency, the reactance of capacitor C_L should be less than $R_E/10$, or less than 160 ohms.

13.3.3 The above circuit has the following frequency and load impedance:-

[1] f_r (frequency) = 80 Hz (approximately)
[2] Z (load impedance) = 7.7 kΩ (approximately)

13.3.4 For the above circuit, the fraction of the voltage that is fed back cannot be worked out because the number of turns and the turns ratio of the coil are not known.

The Armstrong Oscillator

14.1   The Armstrong oscillator is more difficult to design and build.

14.2   Below is a diagram of the Armstrong oscillator:-

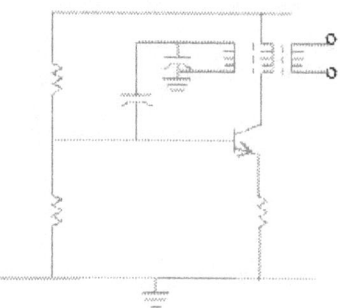

14.2.1   The oscillations of the Armstrong oscillator rely more on the extra winding on the coil than on any other factor.

14.2.2   Having a definite simple procedure for designing an Armstrong oscillator is <u>almost impossible</u> as there is a large variety of transformers and coils available.

14.2.3   Ensuring that coils with the required number of turns are obtained will not be much of a problem, as the number of turns on the coils are specified in the manufacturer's data sheets.

14.2.4   This sees to it that the oscillator will work in its most common operation, at high radio frequencies.

15.1   <u>Remember</u>:
   [1]   To work, an oscillator must have an amplifier, a tuned load and feedback.
   [2]   The oscillator's frequency is determined by the resonant frequency of the tuned load.
   [3]   A voltage divider on the tuned load provides feedback in the oscillator.
   [4]   The three main kinds of oscillator studied so far are the Colpitts, the Hartley and the Armstrong oscillators.
   [5]   Nothing is required to start the oscillations once the circuit has been built; the oscillations should commence spontaneously if the circuit is right.

Successful Oscillator Design

16.1   In building oscillators, selecting the correct coil or inductor is <u>crucial</u>.

16.2   In the laboratory or workshop, it is often hard or impossible to find the exact inductor that is specified in the circuit design.

16.3   Normally, the most readily available coil is utilized, and the rest of the circuit is designed around it.

16.4   Because of this the following four main problems are encountered:-

[1] The inductance value may not be the most suitable one for the frequency range desired.

[2] The exact inductance value may not be known.

[3] Because of capacitance in the coils and the current in the coil windings, et al., the inductance value may vary by a large amount over a narrow frequency range around the desired frequency.

[4] The coil may or may not have tap points or extra windings, which may bring about a change in the original circuit that is desired. (For instance, if there no extra windings an Armstrong oscillator cannot be constructed, and if there are no taps on the coil a Hartley oscillator cannot be built.)

16.5   As the Colpitts oscillator is the easiest of the three to make, it is better to build one.

16.6   When building a Colpitts oscillator, almost any coil can be utilized, given that it is suitable for the frequency range desired, e.g., a coil from the tuner section of a television set is not suitable for a 1 kHz audio oscillator, as the coil's inductance value is outside the range best suited to a low frequency audio circuit.

16.7   The Colpitts oscillator will work over a wide frequency range.

16.8   The following design procedure can be used for building the Colpitts oscillator (and, with some slight modification, the Hartley oscillator):-

[1]   Decide on the oscillator frequency.

[2]   Select a suitable coil. (This presents the greatest practical difficulty, as a coil of the desired inductance value is often not available. In practice, whatever coil is readily available will be used. Luckily, a wide range of inductance values can be utilized.)

[3]   If the inductance value of the coil is known, calculate the capacitor value with the following formula:-

$$f_r = 1/(2\pi\sqrt{LC})$$

(Use this capacitor value for C_1 in the following steps.)

[4]   Select any value of capacitance and call this C_1, if the inductance value is not known. (A frequency which is much different from that required may result. At this point the most important thing is to get the circuit oscillating. The values can be adjusted later.)

[5]   Select capacitor C_2, whose value is to be between three and ten times the value of

C_1. The coil and the two capacitors, C_1 and C_2, are connected as shown below:-

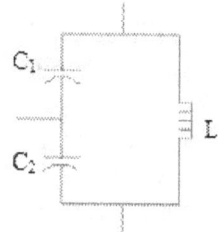

[6] Design an amplifier that has a common emitter gain of about 20. (Select the collector DC voltage to be about half the supply voltage. The value of the collector resistor R_C should be about one-tenth the value of the tuned circuit impedance at the resonant frequency. It is often difficult to choose the collector DC voltage, especially if the value of the coil is not known. Usually, the value of the collector DC voltage is assumed, while R_C is arbitrarily chosen.)

[7] Stop and connect the circuit.

[8] Calculate the value of C_C with the following formula:-

$$C_C = 10^6/(2\pi f_r 160) \times \mu F$$

(Carry out the calculation by choosing X_C to be 160 ohms at the desired frequency, using the "rules of thumb".)

[9] Calculate the value of C_B. (Choose C_B so that X_C is 160 ohms at the desired frequency.)

[10] Apply power to the circuit. Check for oscillations with the oscilloscope. Check the frequency. (Adjust C_1 until the desired frequency is attained, if the frequency is very far from the desired one. To ensure that the ratio of values is about the same, adjust C_2. The output level will be affected by C_2.)

[11] If the circuit does not oscillate, troubleshoot the circuit. (The troubleshooting steps will be brought up shortly.)

[12] Change the feedback connections to the base instead of the emitter.

[13] Calculate C_E. (At the desired frequency, X_C should be 160 ohms. However, X_C should be below 100 ohms if R_E is less than 1 kΩ.)

16.9   Most often the <u>faulty feedback connections</u> cause an oscillator to fail, especially when an unknown coil, which may have several windings or taps, is utilized.

16.10 The following steps can be used to <u>troubleshoot</u> a faulty oscillator:-

[1] Make sure that C_B, C_C and C_E are all large enough to have a reactance value which is less than 160 ohms. Make sure that the value of C_E is less than one-tenth the value of R_E.

[2] Check the C_1/C_2 ratio, which should be between 3:1 and 10:1.

[3] Interchange C_1 and C_2. (They may be in the wrong place.)

[4] Check to ensure that the feedback connections are made to the correct location.

[5] Check to find out whether the feedback connection originates from the correct place.

[6] Check both ends of the tuned circuit to make sure that they are connected to the correct location.

[7] Check the DC voltage level of the collector, the base and the emitter.

[8] Check the values of the capacitors in the tuned circuit. (Randomly change the capacitors until the circuit oscillates, if need be.)

[9] Check to see if any of the components are defective, if all of the above steps fail to bring about oscillations in the circuit. Carefully check the circuit wiring. (It may, e.g., be discovered that the capacitor is shorted, the transistor is dead or its β is too low, and the coil is shorted or opened.)

16.11 The above troubleshooting steps will usually ensure that the oscillator works.

16.12 A working oscillator may have either of or both the following two main defects:-

[1] An output level that is too low. (The sine wave is usually very "pure" and "clean" when this happens. Another transistor should be utilized as an amplifier after the oscillator. For the Colpitts oscillator, the problem will often be rectified by changing the ratio of C_1 and C_2.)

[2] An output waveform which is distorted. (This problem usually arises when the values of C_B, C_C, or C_E are not low enough. This problem may also occur when the oscillator is not working within the best frequency range of the coil, or when there is an output amplitude that is too high.)

11 ABC'S OF ELECTRONIC CIRCUIT DESIGN

1.1   The electronics engineer who designs electronic circuits relies very greatly on data-books from IC and component manufacturers.

1.2   These data-books are normally obtained free-of-charge.

1.3   The designer normally keeps a look-out for the latest ICs available in the market.

1.4   Vendors who supply the ICs would usually keep him posted on the latest available products and technology.

1.5   The rationale is that the company has to keep up with the competition.

1.6   The electronics engineer therefore has to know what the latest technology in the market is - they have to be better than the competition, or, at least, be on the par with them.

1.7   Faster ICs, with more memory capacity, and other state-of-the-art components/parts such as self-resetting power supply cut-off switches (in lieu of fuses) and "8-layer" PCBs, are a boon to the electronics engineer.

1.8   IC data-books show ICs with their related electronic circuits and they provide a host of technical information such as current requirement and input and output voltages.

1.9   Thus, nowadays, the electronics engineer seldom has to design a circuit from scratch as such circuits are freely available for adoption in the data-books.

2.1   However, not all electronic circuits shown in IC data-books would work excellently according to specifications.

2.2   The engineer normally has to make a proto-type of the circuit he has selected and test it out.

2.3   He might have to modify the circuit a little, e.g., by varying the values of a component or two here and there.

2.4   The electronic circuit could only be considered successful, when, on testing, is found to work according to specifications.

2.5   However, a circuit might fail, not because of an inherently faulty design, but due to careless workmanship, such as poor soldering.

3.1   Nowadays, a whole electronic circuit on a PCB could be "compressed" into an IC.

3.2   It pays to look out for ICs which could replace a whole PCB with its many assembled electronic components such as resistors, diodes and capacitors.

3.3   This would help cut down design-time and make the final product more efficient.

3.4   There are organizations which set out to custom-design such ICs.

4.1   Other technologies, e.g., surface-mount-technology, could make the final product smaller, more compact and more efficient.

4.2   Examples of products utilizing some of these latest technologies include lap-top and palm-top computers and laser disc players.

4.2.1	For instance, computers nowadays are not only smaller and take up much less desk space, they are faster in operation and have more interesting features and memory capacities.
5.1	The more ambitious electronic circuit designer might want to design a new circuit from scratch.
5.2	He might have a brilliant concept or two and might spend many long hours and week-ends trying to perfect his very own design (by testing and re-testing his proto-type and checking and rechecking his calculations).
5.3	If his design is found to work out well, he could end up with a patent for it.
6.1	In short, the electronics engineer should always be alert and on the look-out for the newest devices and components/parts that are available in the market.
6.2	He should try to cultivate a good working relationship with the vendors of devices and components.
6.2.1	He should be able to tap their knowledge and expertise.
6.3	He should be curious and keen to learn new things.
6.4	He should maintain a good library of component data-books and other reference books.
6.5	The component data-books would offer a wide range of electronic circuits, e.g., audio, video and transistor circuits, for him to choose.
6.5.1	For instance, different semi-conductor manufacturers offer various devices with differing electronic characteristics and values and the indicated electronic circuits incorporating these devices also vary.
6.6	The electronics engineer has therefore to use his discretion, based on his experience and knowledge and on the cost/benefit factor, to select the device and electronic circuit that would deem to best suit his requirement.

12 SOME PERSONAL TIPS ON CIRCUIT DESIGN

1.1   Proficiency in circuit design would come with experience.

1.2   The author would like to provide some pointers to the budding circuit designer, which are as follows:-

[1]   Keep a proper documentation (make proper notes) of every stage of the design. Any changes in the design must be documented. This facilitates fault-tracing when things do not turn out right.

[2]   Keep a note-book (for reference) of useful circuits (which you have designed and used effectively in the past). Always refer back to this note-book when working on the design.

[3]   Have a well-stocked library of integrated circuits (ICs) data-books from the various semiconductor manufacturers, e.g., Texas Instrument, NS, Matsushita, et. al., as well as data-books on other electronic parts such as diodes, capacitors, transformers, et al. You might have to be selective of the parts to be used and such data-books could provide a wealth of useful technical information. For example, a whole lot of electronic parts in a certain stage of your design might be substituted by a simple IC, after carefully studying the data of the selected IC in the data-book.

[4]   Always keep an open mind. No two circuit designers would come up with the same design.

[5]   Try to minimize on utilization of components, parts for cost-saving, time-saving and design efficiency purposes.

[6]   Maintain a library of reference books on electronic circuits, for easy and quick reference, as well as electronics journals with electronic circuit ideas.

[7]   Intermingling and discussions with fellow professionals, during professional society activities and/or seminars/conferences, would help boost circuit design creativity.

[8]   Always try to keep up-to-date by reading widely, attending courses and talks, visiting trade shows and exhibitions, et al.

[9]   Remember that in a dynamic specialization such as electronics it is necessary to be prepared for life-long learning. (Thus, you should have developed your learning skills.)

[10]   Nowadays, advanced software such as Orcad and Mentor Graphics allow you to design, simulate and test the circuit (you have designed) by the computer alone (which means that you do not have to construct a physical model based on your design to test (by using the multimeter, oscilloscope, logic probe and what-not)). Therefore, it is important to keep up-to-

date with computer technology and software.

1.3   The author would like to assure the reader that with experience, enthusiasm for learning and exposure a good circuit designer is likely to result.

13 EPILOGUE

1.1   Modern electronics can be split into two main branches: analog electronics and digital electronics.

1.2   The whole field of digital electronics depends on the principle of transistor switching.

1.3   Digital circuits are just several simple transistor switching circuits that are used many times over, e.g., those found in a computer.

1.4   The basis of all analog or linear circuits is the transistor amplifier. (Most analog circuits are made up of several transistor amplifiers, and are currently based on the IC op-amp (operational amplifier).

1.5   The other aspect of electronics that has been touched on here is related to the non-amplifying electronic components which are called passive components.

BIBLIOGRAPHY

1) Horowitz and Hill. The Art Of Electronics. Cambridge University Press. 1980.
2) Jones. A Practical Introduction To Electronic Circuits. Cambridge University Press. 1981.
3) Melville. Electricity. The Hamlyn Publishing Group Limited. 1978.
4) Worcester. Electronics. The Hamlyn Publishing Group Limited. 1980.
5) Duncan. Adventures With Electronics. John Murray. 1978.
6) Duncan. Adventures With Microelectronics. John Murray. 1981.
7) Duncan. Adventures With Digital Electronics. John Murray. 1982.
8) Sinclair. Practical Electronics Handbook. Newnes Technical. 1980.
9) Pierce and Paulus. Applied Electronics. Merrill. 1972.